FRESHWATER SPONGES OF TENNESSEE

FRESHWATER SPONGES OF TENNESSEE

John Copeland • Stan Kunigelis

THE UNIVERSITY OF TENNESSEE PRESS
Knoxville

Copyright © 2025 by The University of Tennessee Press / Knoxville.
All Rights Reserved. Manufactured in the United States of America.
First Edition.

LIBRARY OF CONGRESS CATALOGING-IN-PUBLICATION DATA

Names: Copeland, John, author. | Kunigelis, Stan, author.

Title: Freshwater sponges of Tennessee / John Copeland, Stan Kunigelis.

Description: Knoxville : The University of Tennessee Press, [2025] | Includes bibliographical references and index. |

Summary: "This book provides Tennesseans and a larger professional and lay audience an introduction to freshwater sponges, their biology, natural history, and identification. Authors John Copeland and Stan Kunigelis argue that freshwater sponges, while studied less frequently than their marine sponge counterparts, are far more accessible to both professional and amateur biologists given the plethora of rivers, particularly here in Tennessee. The authors discuss the natural history of sponges, watershed health and environment, sponge anatomy, taxonomy and classification, and the specific species students, researchers, and readers are likely to find in Tennessee's river system. The book is also replete with color photos from the field and the lab alongside reproductions of microscopic slides detailing Tennessee's freshwater sponges"
— Provided by publisher.

Identifiers: LCCN 2024046661 (print) | LCCN 2024046662 (ebook) | ISBN 9781621909194 (paperback) | ISBN 9781621909200 (kindle edition) | ISBN 9781621909217 (adobe pdf)

Subjects: LCSH: Freshwater sponges—Tennessee.

Classification: LCC QL373.D4 C67 2025 (print) | LCC QL373.D4 (ebook) | DDC 593.4/609768—dc23/eng/20241016

LC record available at https://lccn.loc.gov/2024046661
LC ebook record available at https://lccn.loc.gov/2024046662

JOHN E. COPELAND dedicates this book
to Mary Copeland, Luke Copeland,
Lydia Copeland Gwyn, Thomas Gwyn,
Seamus Gwyn, and Vivian Gwyn,
all of whom have taught and inspired me.

STAN C. KUNIGELIS dedicates this book
to four generations of family microscopists,
Stan Sr., Kate, Heather, and Ernestine
who strive to discover the world around them
and heal their brothers and sisters in need.

CONTENTS

	Preface	ix
	Acknowledgments	xiii
	Introduction	1
ONE	Why Freshwater Sponges?	7
TWO	Anatomy	23
THREE	Natural History	37
FOUR	Field and Laboratory Techniques	61
FIVE	Taxonomy, Classification, Identification	65
SIX	Sponges of Tennessee	71
SEVEN	Environmental Variables and Organismal Interactions	147
EIGHT	Conservation	161
	Appendix A. Suggestions for Teachers and Researchers	169
	Appendix B. Distributions of Freshwater Sponge Species in the United States	171
	Glossary	187
	Selected References	195
	Index to Species	211

PREFACE

As biologists we have the privilege of working in a region that is a wonderland, the Cumberland Gap area of Tennessee, Kentucky, and Virginia. We are minutes away from some of the most biologically diverse rivers of the temperate regions of the world. The southeastern region of the United States is known for its biodiversity. This is especially true for the Appalachian regions of Kentucky, North Carolina, Tennessee, and Virginia. The aquatic biota of this region has levels of diversity and endemism rivaling those of tropical regions. Unfortunately, this biological wonderland is changing at an accelerating rate, faster than we ever thought it would.

Aldo Leopold, the father of wildlife ecology, wrote: "To keep every cog and wheel is the first precaution of intelligent tinkering."[1] With this book we seek to inform the reader of some cogs and wheels that are often overlooked by biologists and are unknown to many people. We are losing cogs and wheels at an alarming rate. Biologists have documented that the Earth is now in its sixth mass extinction and have determined the loss of biodiversity in aquatic environments to be far greater than that for terrestrial ecosystems. It is thought that over 80 percent of Earth's life-forms have not been discovered. Extinction rate estimates suggest that current rates are 1,000 to 10,000 times higher than historical natural background rates. Statistics reveal we are losing species before they are discovered and described. Should we not know what we have before we lose it, hoping not to face the loss?

Modifications to some Tennessee drainages have occurred, and not without consequences. Construction of dams, channelization of streams, industrial and agricultural pollution, strip-mining acidification, and sedimentation from lands under development or alteration have certainly altered aquatic environments. For many aquatic species we know the results of the consequences. Such modifications on Tennessee's freshwater sponges are unknown because historical data are lacking.

---

1. Leopold, Aldo. *Round River: From the Journals of Aldo Leopold.* (1953; repr., New York: Oxford University Press, 1993), 146.

Sponges were first documented in Tennessee in 1943. Only five species of sponges had been reported from Tennessee by 2012. As of 2023, fourteen freshwater sponge species are known to occur in Tennessee waters. The current information regarding Tennessee's freshwater sponges only scratches the surface of that needed by conservation agencies to manage this fauna.

As biologists we realize freshwater sponges fulfill critical roles necessary to the functioning of aquatic ecosystems. Unfortunately, there are some who fail to understand the importance of our natural heritage. We have been asked, "What good is a freshwater sponge?" Allow us to quote Aldo Leopold one more time: "The last word in ignorance is the man who says of an animal or plant, what good is it? If the land mechanism as a whole is good, then every part is good, whether we understand it or not."[2] As stewards of planet Earth, we are charged to manage its resources wisely, with care, concern, proper use, and when necessary, protection.

From an egocentric perspective, we often ignore that which is not directly relevant to our needs, distracting us from the unified beauty and significance of each ecological puzzle piece, with each unique piece being integral to a balanced ecology.

As we write this preface, the natural resources and lands held in trust for all citizens and future generations are under attack due to anthropogenic causes. The belief that nature can be treated as a resource to be plundered and exploited for personal gain is not only selfish but escalates a degenerative spiral of denial to future generations. This is a failed stewardship philosophy. Have those of us who have a love and an appreciation for nature failed to do an adequate job of educating the public about the importance of biodiversity and maintaining natural places and wild things? Whether the current situation results from a lack of education or denial of responsibility, our goal is to begin the healing process.

Every discipline has its own unique language. Mechanics speak of manifolds, flywheels, carburetors, and universal joints. Lawyers speak of subpoenas, stare decisis, amicus curiae, and de novo. Biologists have their own lexicon. Biology students are often intimidated by biological jargon. Fortunately, as they continue their biological studies, they begin to pick it up and use it. It is not an easy task to take the complex vocabulary of a specific field of biology and convert

---

2. Leopold, *Round River*, 146.

it into language easily understood by non-biologists. To assist the reader, we provide definitions in the text where we think appropriate; otherwise, definitions are found in the glossary.

We have written a science book with the layperson in mind for several reasons, including:

> We wish to share our excitement and fascination about a group of organisms most people either know little about or do not know exist.
>
> We wish to share what we have learned with a broader audience than the scientific community.
>
> We know the conservation and management of Tennessee freshwater sponges can be advanced through educating those unaware of this natural resource.
>
> We encourage high school teachers, college professors, professional biologists working for private and government agencies, and laypersons with an interest in natural history to consider advancing the knowledge of freshwater sponges through research and teaching.
>
> The public will make the final decisions concerning the biological issues our society faces. To make informed decisions, the public should have a knowledge and understanding of the biological issues we currently face, such as the loss of biological diversity and the sixth mass extinction, the interconnectedness of every part of the natural world, climate change, and others.
>
> We are educators who recognize an unfilled niche.

With this book we introduce readers to freshwater sponges by providing information on 14 of the 37 freshwater sponge species reported from North American waters. We believe this book will be informative for laypersons, biology teachers, naturalists, and professional biologists. Topics covered include anatomy, ecology, and the natural history of freshwater sponges. Additionally, we provide information on collecting sponges, preparing sponge material for viewing microscopically, identification of the sponges of Tennessee, and a few suggestions for teachers interested in studying freshwater sponges.

In most invertebrate biology classes freshwater sponges are overlooked in favor of marine sponges. Instructors having no practical experience with freshwater sponges typically inform students of their existence after which they spend the rest of the class time discussing marine sponges. Additionally, most invertebrate biology courses taught at inland universities do not take field trips to the ocean while freshwater habitats supporting freshwater sponges exist only a few minutes from campus. Trips to local streams can result in field time and a collection of sponges appropriate for teaching aspects of microscopy, taxonomy, biological diversity, ecology, natural history, sexual and asexual reproduction, metamorphosis, totipotent or stem cell concepts, symbiosis, dispersal, food webs, interactions between and among different species, and competition among species. Appendix A provides additional suggestions for teachers and researchers.

This body of work on the freshwater sponges of Tennessee is a result of the encouragement of the late Dr. Louis Lutz of Lincoln Memorial University. While working on *Io fluvialis* Say, 1825, the spiny river snail, in the Clinch and Powell Rivers, Lutz observed that freshwater sponges were abundant. Lutz on several occasions encouraged one of us, Copeland, to "take a look" at freshwater sponges. After the passing of Dr. Lutz, Copeland began the work which led to the publication of this book. We dedicate this book in memory of our late friend and colleague, Dr. Louis Lutz.

ACKNOWLEDGMENTS

For several years we have worked to advance our knowledge concerning Tennessee's freshwater sponges. This work could not have occurred without the assistance of several individuals. Many Lincoln Memorial University professors and students freely gave of their time. We especially acknowledge Jessie Tussing, Tucker Jett, Chase Rich, Emily Stewart, and Kayleigh Hanson for the large amount of time they spent in the field collecting sponges and processing samples for microscopic examination.

To Rizwan Ahmed, Mustafa Ali, Patty Bottari, Ron Caldwell, Mary Copeland, Dedra Erwin, Steve Furches, Lydia Gwyn, Thomas Gwyn, Seamus Gwyn, Vivian Gwyn, Vicky Inger, Genevieve Kemp, Dennis Kiick, Sammy Knefati, Victoria Long, Jared Miller, Dan Oakley, Daniel O'Bryan, Gayatri Ravi, Kate Regan, Bill Reeves, Adam Rollins, Maggie Singleton, Capt. Vic Scoggins, Chelsie Smith, Jessica Stegner, Aggy Vanderpool, Jennifer Watson, Stephanie Williams, and David Withers we express our gratitude for their contributions and their assistance in field collections or laboratory processing of sponges.

We gratefully acknowledge Scott Nichols, Denver University, for permission to use the scanning electron microphotograph of a choanocyte chamber (Fig. 14) and Mike Chadwell for permission to use his outstanding illustrations (Figs. 10, 15, 16, 17, 18, 19, 20, 21, and 23).

We thank the United States Geological Survey for providing the map used to show sponge distributions, and NatureServe Network for the map of imperiled aquatic invertebrate species (Fig. 5). The authors acknowledge Oxford Publishing Limited for graciously allowing the use of quotes of Aldo Leopold taken from his work Round River.

We thank the Tennessee Wildlife Resources Agency, Lincoln Memorial University, and the Well Being Foundation for providing funding to complete the field and laboratory aspects of this project.

The information presented in this book is the result of the research of many biologists, past and present, from around the world. We thank them for teaching us. John Copeland greatly acknowledges Roberto Pronzato of Dipartimento di

Scienze della Terra, dell'Ambiente e della Vita Università di Genova, and Renata Manconi of Dipartimento di Scienze della Natura e del Territorio, Università di Sassari, for their international cooperation with describing and naming of *Cherokeesia armata* and *Heterorotula lucasi*. Their extensive body of published works is essential reading for anyone interested in freshwater sponges.

Introduction

"Declare the past, diagnose the present, foretell the future."

—HIPPOCRATES—

Do freshwater sponges really exist? Why would anyone want to study freshwater sponges? These are just two of the many questions our friends asked when we informed them we study freshwater sponges.

Yes! Freshwater sponges really do exist and have for millions of years. The oldest undisputed freshwater sponge fossil dates to the Permo-Carboniferous time.

Biologically, the Carboniferous Period was a time of firsts. The first true bony fish, the first sharks, the first amphibians, and the first reptiles all arose during this period. The Carboniferous landscape was one of extensive swamp forests dominated by lycopods, sphenopsids, cordaites, seed ferns, and true ferns. Seed ferns were the first plants to produce seeds. Current living lycopods and sphenopsids are diminutive in size compared to their Carboniferous ancestors. Lycopods are represented in the modern world by club mosses, but in the Carboniferous Period they included *Lepidodendron* and *Sigillaria* which grew to be about 100 feet tall. Modern sphenopsids include *Equisetum* (horsetails) which may grow to be 4 to 5 feet tall but during the Carboniferous Period the giant horsetail, *Calamites*, grew to heights of 60 feet. These forests became the coal beds that characterize the Carboniferous stratigraphy.

A major reproductive development occurred during the Carboniferous Period. Reptiles produced the first amniotic eggs. The development of the amniotic egg was a major factor in opening dry environments for occupation. Eggs no longer needed to be deposited in water for the development of the embryo and larva. The shell of reptilian-produced eggs provides protection to the developing embryo, allows for water retention, and is permeable enough for the exchange of gases with the atmosphere.

Reptiles, birds, and mammals are amniotes. Amniotes are vertebrate organisms that produce a membranous sac, the amnion, which encloses the developing fetus. In reptiles and birds, the amnion is found within the shell-encased egg, while in mammals it encloses the fetus within the uterus.

Carboniferous freshwater sponges shared their freshwater environments with two important fish groups. Both groups were osteichthyans (fish having bony skeletons) represented by the classes Sarcopterygii (lobe-finned) and Actinopterygii (ray-finned fishes). The Sarcopterygians are important for giving rise to tetrapods (animals having four feet). In the modern world, lobe-finned fishes are a small group consisting of lungfishes and the coelacanth. In contrast, the ray-finned fishes comprise over fifty percent of all living vertebrate species.

If you have spent time wading, swimming, or snorkeling in streams and rivers, then you have probably seen the descendants of Carboniferous freshwater sponges. You may have thought you were viewing a mat of algae or a plant. Occasionally, they are small and inconspicuous, but they can be so large they can't be overlooked.

Humans have a long history of using marine sponges. For thousands of years humans have used marine sponges for drinking, bathing, scrubbing, absorbing, collecting, transferring, and removing water from where it is not wanted. The uses made of marine sponges are practical and continue today. Freshwater sponges are not useful in these ways.

Commercially used marine sponges of the genera *Spongia* and *Hippospongia* do not have a mineral skeleton whereas freshwater sponges have a hard-mineralized skeleton. Also, freshwater sponges normally grow as thin encrustations or cushions on their substrate, while marine sponges with commercial uses have bodies that are larger, thicker, upright, and bulbous.

If freshwater sponges are not used for bathing, cleaning, and absorbing spilled liquids, then what uses could they have? The fact that some species of bacteria and fungi have become resistant to current antibiotics is not lost on the public. Who has not read about or listened to accounts of "flesh-eating bacteria?" Necrotizing fasciitis is the disease caused by such bacteria. This is a horrific disease that spreads quickly and aggressively resulting in tissue death. Group A *Streptococcus* (GAS) and some species of *Staphylococcus* are known causative pathogens of this disease. Unfortunately, some strains of *Streptococcus* and *Staphylococcus* have become resistant to antibiotics used in their treatment. The number of species of bacteria that have developed resistance to conventional antibiotics is growing, resulting in a serious medical problem.

Scientists are constantly looking for sources of new antibiotics. In the past few years, sponges have been found to produce a variety of defense compounds, some of which are antibiotics. Compounds produced by the freshwater sponge *Orchridaspongia rotunda* Arndt, 1937 (sorry about the use of scientific names but most freshwater sponges have not been assigned common names) have been found to have antimicrobial activity. A methanol extract and an acetone extract were found to be quite effective against the bacteria and fungi studied. The methanol extract proved to be more effective against bacteria than the control antibiotics, streptomycin and ampicillin, and the acetone extract was more effective against fungi than the controls bifonazole and ketoconazole. This field of inquiry holds a great deal of promise and is of considerable interest to the pharmaceutical industry.

For at least 300 years Russian, Polish, and Ukrainian physicians have used a freshwater sponge known as Badiaga for treating patients having lung disease, rheumatism, and sores. Badiaga is dried, ground into a powder, and rubbed on the chests and backs of patients with lung problems. In the 1930s, it was discovered that Badiaga was not a single sponge, but rather a mixture of several freshwater sponges including *Spongilla lacustris* (Linnaeus, 1759), *Eunapius fragilis* (Leidy, 1851), *Ephydatia fluviatilis* (Linnaeus, 1759), and *Ephydatia muelleri* (Lieberkühn, 1856). Badiaga can be purchased from sellers of homeopathic medicines and online where it is sold as Badiaga Spongilla powder.

Spicules of freshwater sponges have occasionally caused human health problems. In Brazil, skin and mucus membrane lesions occurred after swimming in waters with high concentrations of spicules. Lesions have manifested in some people as pruritic erythematous papules (a raised skin rash). In an analogous manner, Indigenous Brazilians using spicules in the production of ceramics developed hand lesions. Treatment involved corticosteroids such as the synthetic, anti-inflammatory glucocorticoid, prednisone. A more severe situation involving eye lesions occurred in Brazil. Spicules encountered while bathing, swimming, and diving with eyes open caused immediate eye irritation and itching followed later by eye lesions. Some cases resulted in severe visual impairment.

Freshwater sponge spicules are currently used in the production of some human skin care products. *Spongilla* spicules, in either powders or gels, are used as exfoliates for removing surface skin cells. Promotions and advertisements report sponge-containing cosmetics to have several properties important for maintaining healthy skin, such as stimulating the dermal layer of skin, activating skin microcirculation, opening blocked pores, promoting collagen production, and treating acne. In their pursuit of a reliable supply of spicules (and control over product quality thereby), companies have developed methods for cultivating sponges.

Several freshwater sponge cells are totipotent cells. Totipotent cells are stem cells. There are unique differences between sponge totipotent cells and those of other organisms. These differences could prove beneficial to research on human stem cells.

In regions of North America, the Brazilian Amazon, and elsewhere, Indigenous peoples have used freshwater sponge spicules as a tempering agent in the production of ceramics. Temper is any material that when added to clay fortifies it for the firing process, making items made from clay less likely to break when used. Without the use of microscopes to see spicules, how did Indigenous peoples know to use spicules as temper? How were spicules acquired? Older ethnographic studies in the Amazon Basin showed that Indigenous ceramic communities collected, dried, and burned sponges. More recent archaeological work indicates Amazonian potters to have used spicule-rich clay sources formed by decomposing sponges. Which is correct? A study using scanning electron microscopy (SEM) of pottery shards detected the presence of sponge gemmules (asexual reproductive structures). Because gemmules rarely survive sedimentation researchers concluded that Amazonian potters manually added spicules to their clay. However, it may be determined from future studies at other locations that some clays are naturally abundant in spicules due to decomposing sponges.

Freshwater sponge spicules have been useful, as bio-indicators, for determining environmental and hydrological phases in paleoenvironments. A study concerned with determining the hydrological phases of the Parana River hydrological region in the Taquarussu Region of Brazil from late Pleistocene to mid-Holocene used spicules of freshwater sponges to determine a sequence of three phases. Finding no spicules in the oldest sediments, spicules of river-dwelling sponges in younger sediments, and bog-dwelling sponges in the youngest sediments led researchers to suggest a three-phase evolution from dry land to a fluvial environment to a lacustrine environment.

The ecological roles freshwater sponges perform in their ecosystems are more important than any benefit they provide humans. All organisms of a biological community contribute to its operation. Some, such as keystone species, are more important than others, but all contribute. Freshwater sponges contribute in important ways, but we know of no ecosystem in which they are the keystone species. Nevertheless, they perform roles and occupy a niche no other organism can.

ONE
Why Freshwater Sponges?

"The world as we have created is a process of our thinking. It cannot be changed without changing our thinking."
—ALBERT EINSTEIN—

Freshwater sponges? Really? That is a frequent reaction when people are informed of the existence of freshwater sponges. Even though sponges are common in freshwater environments they remain unknown to most people. Freshwater sponges have colonized not only rivers and lakes, but also human constructed ponds, water tanks, pipelines, and sewage treatment facilities. Sponges are found from the water's surface to depths hundreds of meters below the surface. Even though sponges are simple organisms they are one of the most biologically successful life-forms to have arisen on Earth. Sponges have been members of aquatic ecosystems for over 600 million years and survived events that resulted in the extinction of other organisms.

Initially, biologists found it difficult to determine whether freshwater sponges were plants or animals, or a mixture of the two. In physical appearance they appear as cushions or encrustations (Fig. 1) on the surface of substrates but can be branched and fingerlike. Colors range from white to buff, yellow to brown, gray to black to green (Fig. 2).

When first discovered, sponges were thought to be plants. Perhaps because sponges are sessile (nonmotile), some grow in a branchlike manner, and some are brown or green in appearance, all of which are plantlike characteristics. A

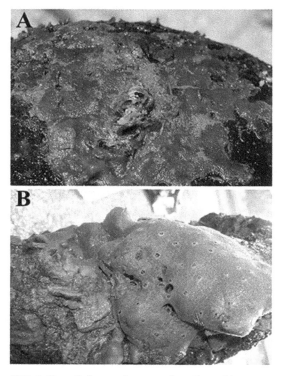

FIG. 1. Growth forms. **A,** encrusting; **B,** cushion.

specimen of the freshwater sponge *Spongilla lacustris* (Linnaeus, 1759) has been found in the herbarium of Carolus Linnaeus. An herbarium is a collection of preserved plant specimens. Linnaeus, who is known as the father of modern taxonomy, was a Swedish botanist, physician, and zoologist who formalized binomial nomenclature, the current system of naming organisms.

Sponges are not plants! Sponges were recognized as animals in 1765 when their ability to generate an internal water current was described. Freshwater sponges belong to a diverse group of animals placed in the Phylum Porifera. Porifera means to be porous or to have pores. The body of a sponge is covered with pores through which water enters and exits the sponge. Sponges obtain their nutrition by filtering foods from water passing through their bodies. Think of a sponge as a living water pump. Among sponges, freshwater sponges are a minority group making up less than 3% of the total number of known species.

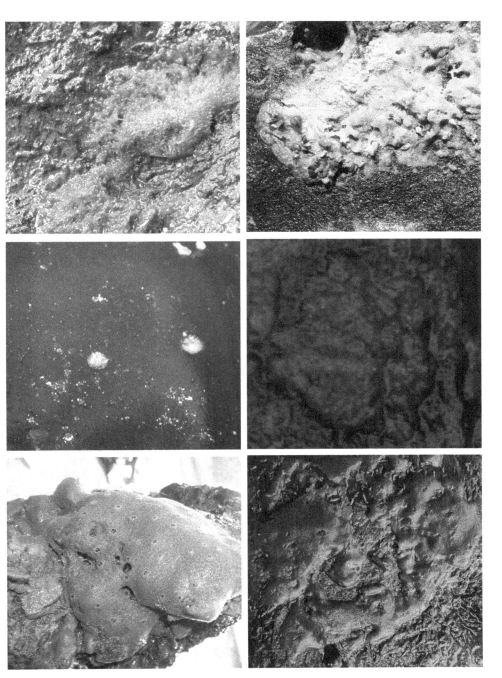

FIG. 2. Some colors of freshwater sponges.

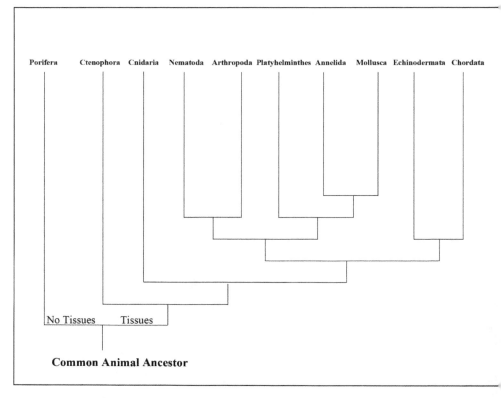

FIG. 3. Animal phylogenetic tree showing the sister group relationship of Porifera (sponges) to other animals. Porifera (sponges), Ctenophora (comb jellies), Cnidaria (jellyfish, corals), Nematoda (round worms), Arthropoda (insects, crustaceans, spiders), Platyhelminthes (flat worms such as a tape worm), Annelida (earthworm), Mollusca (snails, mussels), Echinodermata (starfish), Chordata (vertebrate animals: amphibians, reptiles, birds, mammals).

Sponges are invertebrate animals differing from other animals by lacking true tissues and organs and having asymmetrical bodies. Instead of tissues and organs, sponges have specialized cells that perform all the functions necessary for their survival. Because sponges (sub-kingdom Parazoa) and all other animals (sub-kingdom Metazoa) have the same common ancestor, they form a sister group. Sister groups are groupings of organisms that not only have the

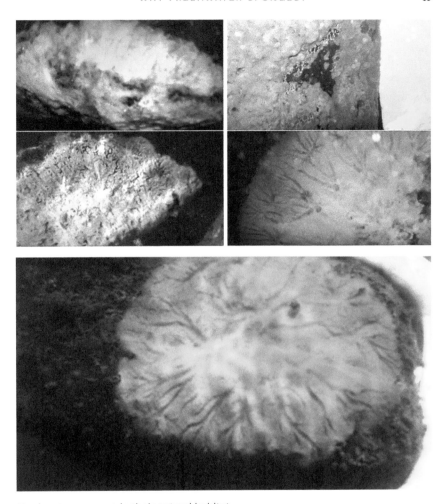

Freshwater sponges in their natural habitat.

same immediate common ancestor but are each other's closest relative (Fig. 3). This sister grouping was considered settled science. However, within the past few years some biologists have presented data suggesting ctenophores (comb jellies) as the sister group to all other animals. An ongoing debate arose between those favoring sponges as the sister group and those favoring ctenophores. Recent studies contradict one another, so the debate continues.

A major difference between plants and sponges is how they acquire energy. Plants are autotrophs, which use solar energy to make complex organic food molecules (carbohydrates, lipids, and proteins) from the inorganic carbon compound carbon dioxide. In contrast, sponges are heterotrophs, which consume other organisms to obtain organic food molecules. Energy held by the chemical bonds of carbohydrate molecules, and when needed lipids and proteins, is used to produce adenosine triphosphate (ATP) molecules. ATP molecules provide the energy needed by organisms to live.

Some sponges appear green due to the presence of endosymbiotic algae. A common misconception is plants do not utilize the green wavelength of light for photosynthesis. The truth is most of the green light is useful in photosynthesis, but a small amount is reflected. Plants reflect more of the green wavelength and use it less effectively than any other wavelength. Green sponges are green because of the small amount of green light reflected by their algae endosymbionts.

Sponges are considered by many biologists to be the most primitive animal living on Earth and the first multicellular animal to have appeared in ancient oceans of the Ediacaran Period (635–541 MYA). The Ediacaran Period is the last period of the Neoproterozoic Eon (Fig. 4). Unfortunately, the fossil record of freshwater sponges is poor and discontinuous. Soft bodied organisms, such as sponges, normally do not fossilize because, lacking a protective hard outer layer, they decompose quickly after death. Fortunately, spicules of freshwater sponges do fossilize and have proven useful for differentiating species. The use of "molecular fossils" or "biosignatures" has proven useful for providing evidence of once-living life-forms in rock strata devoid of typical fossils. The molecular fossil 24-isopropylcholestane, a diagenetic product of the sterol 24-isopropylcholesterol, is accepted as a "sponge biomarker" and has been found in rocks over 600 million years old. This finding suggests sponges were present in the seas of the Ediacaran Period, over 60 million years before an explosion of new life-forms occurred during the Cambrian Period (541–485 MYA). The identification of Ediacaran fossils is controversial because they suggest animals appeared millions of years earlier than previously thought. The consensus is that representatives of the phyla Porifera, Cnidaria, and Ctenophora were present in Ediacaran seas. Some scientists think it probable that members of the phyla Mollusca, Annelida, Arthropoda, and Echinodermata were also present.

EON	ERA	PERIOD	EPOCH	MYA
P H A N E R O Z O I C	Cenozoic	Quaternary	Holocene	0.012–Present
			Pleistocene	2.58–0.0012
		Neogene	Pliocene	5.33–2.58
			Miocene	23–5.33
		Paleocene	Oligocene	34–23
			Eocene	56–34
			Paleocene	66–58
	Mesozoic	Cretaceous		145–66
		Jurassic		201–145
		Triassic		252–201
	Paleozoic	Permian		299–252
		Carboniferous (Miss. & Penn.)		359–299
		Devonian		418–359
		Silurian		444–419
		Ordovician		585–444
		Cambrian		541–485
P R E C A B R I A	P R O T E R O Z O I C	Neoproterozoic	Ediacaran	635–541
			Cryogenian	720–635
			Tonin	1000–720
		Mesoproterozoic	Stenian	1200–1000
			Ectasian	1400–1200
			Calymmian	1600–1400
		Paleoproterozoic	Statherian	1800–1600
			Orosirian	2050–1800
			Rhyacian	2300–2050
			Siderian	2500–2300
	AECHEON			4000–2500
	HADEAN			> 4000

FIG. 4. Geologic time scale.

Fossils from the Cambrian explosion indicate this was a time of rapid animal diversification during which most animal phyla appeared. Fossil representatives of the phyla listed previously are present without question during the Cambrian Period. This explosion of new life-forms lasted only about 20 to 25 million years, which is a brief period of geological time. Current paleontological data suggest the most ancient freshwater sponges date to the Permo-Carboniferous Period. The oldest freshwater sponge fossils reported from North America date to the upper Jurassic Period (201–145 MYA). Most freshwater sponge fossils discovered have been from the more recent Eocene to Pleistocene Epochs.

Controversy surrounds a fossil dated to 890 MYA. This fossil was discovered at Little Dal reefs, Stone Knife Formation of northwestern Canada. It resembles the spongin fiber network of modern-day keratose sponges. If this fossil could be confirmed to be that of a sponge, it would provide physical evidence animals existed over 300 million years earlier than the fossil record and biosignatures indicate.

It has been speculated that during the Permian Period of the Paleozoic Era (250–300 MYA), gemmule-producing sponges inhabiting shallow marine and intertidal habitats were exposed to changing water salinities. Over time sponges adapted to waters with lower salinity concentrations, eventually allowing sponges to adapt to and invade fresh waters. Freshwater sponges have successfully colonized lakes and rivers, and are now commonly found in these ecosystems. They developed specific adaptations to colonize freshwater environments, including the ability to survive extreme temperature ranges, long-term desiccation, and anoxia. Since the beginning of the Industrial Revolution and modern-day agricultural practices, freshwater sponges have adapted to various pollutants.

The movement from salt water to fresh water was probably not an easy transition. Freshwater environments presented difficult problems that marine sponges had to solve, differences in salinity being a major one. The average salt concentration of seawater is about 3.5% (about 35 grams of salt per liter), whereas average salinity concentration in fresh water is about 0.1%. This difference is huge and forced sponges attempting to invade fresh water to overcome osmoregulation problems.

Issues concerning desiccation and dispersal had to be resolved. Ocean environments are continuous with little probability of becoming dry land. In contrast, freshwater environments will have low water levels or a complete dry-up because of insufficient rainfall. Additionally, freshwater ecosystems have

Buffalo River, Lewis County, Tennessee.

a discontinuous spatial distribution that makes it more difficult for sponges to disperse from one body of water to another.

Desiccation and dispersal problems were solved through the development of cryptobiotic resting bodies known as gemmules. Gemmules are small round or spherical structures that contain totipotent cells capable of becoming a sponge. Gemmules are especially important in the life cycle and survival of freshwater sponges. Their presence in the fossil remains of *Palaeospongilla chubutensis* Ott and Volkheimer, 1972, indicate that gemmules have been produced by sponges since at least the Cretaceous Period.

Although invasion of fresh water by sponges occurred by at least the late Carboniferous Period, only six fresh water families are known worldwide. The number of extant sponge species is approximately 8,850, of which only about 250 to 260 or about 3% are freshwater sponges.

When sponges colonized fresh waters, they became members of a complex and diverse community of aquatic organisms. Biodiversity is defined as the

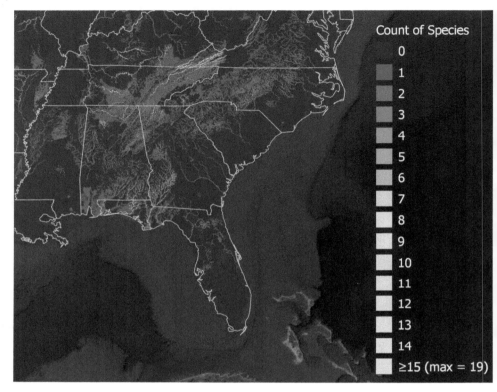

FIG. 5. Distribution map of species richness for imperiled aquatic invertebrates. Shows the importance of the Tennessee River and its headwater tributaries (Clinch, Holston, and Powell Rivers) for providing critical habitat for rare, threatened, and endangered aquatic species.

variety and variability of life-forms comprising a community, ecosystem, biome, or biogeographical realm. A measure of this variety is species richness, which is a count of the number of species in a specified area. However, biodiversity is much more than the count of species found in an area. It is a measure of variation at the genetic, species, and ecosystem levels.

Biodiversity is important for several reasons. Ecologically it is critical for maintaining the balance of ecosystems from the standpoint of nutrient storage and recycling, and protecting soils and aquatic resources. Biodiversity is

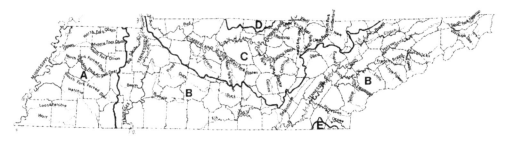

FIG. 6. Major river drainages of Tennessee. **A**, Mississippi; **B**, Tennessee; **C**, Cumberland; **D**, Ohio; **E**, Conasauga. From Kernodle, 1972.

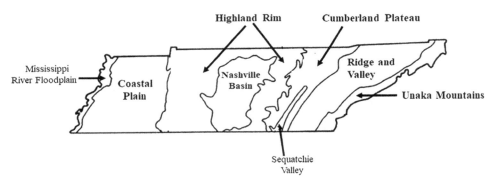

FIG. 7. Some physiographic regions of Tennessee with six provinces in bold font. After Luther, 1977.

beneficial to humans by providing food, medicines, fiber products, pollination of plants, and oxygen in the atmosphere.

The central and southern Appalachian Mountains comprise one of the most biologically diverse temperate regions on Earth. The extremely diverse topography of this region has resulted in microhabitats necessary for the survival of several relict species, some of which have their distribution limited to a single watershed. This is a perilous situation as extinction is a real possibility.

The Tennessee River and its headwater tributaries, the Clinch, Holston, and Powell Rivers, provide critical habitat for several imperiled species.

FIG. 8. Terrestrial Biogeographic Realms. Red line approximate boundaries.

Imperiled species are animals and plants with declining populations that may be in danger of extinction. Since it harbors more imperiled species than any other large river basin, the Nature Conservancy considers the Tennessee River basin to be the most biologically diverse river system for aquatic organisms in North America (Fig. 5). Over 325 species of fish have been documented in Tennessee waters. With at least 78 species of crayfish, Tennessee is home to more species of crayfish than any other state. Only Alabama has more species of mussels. Tennessee ranks fourth in the United States for species of aquatic snails. Approximately four hundred species of caddisflies have been collected from Tennessee's rivers and streams. This remarkable diversity is due to the state's unique hydrology, geology, and physiographic provinces. Five major river drainages and six physiographic provinces are found within Tennessee (Fig. 6 and Fig. 7). The Clinch, Elk, and Powell Rivers are known for their diverse fish and freshwater mussel communities. The Duck River, due in part to the large numbers of fish, freshwater mussels, and aquatic snail species found in it, is considered the most biologically diverse river within the United States. Unfortunately, the contribution freshwater sponges make to aquatic biodiversity is unknown for many of Tennessee's lakes, rivers, and streams, simply because their presence has not been documented.

Biologists have divided the Earth into biogeographical realms, which represent the broadest divisions of the Earth's land surface. Realms are based on distributional patterns of terrestrial organisms and divided into biomes. The diversity of freshwater sponges at the biogeographical scale ranked from highest to lowest is Neotropical, Palearctic, Afrotropical, Oriental, Australasian, Nearctic, and Pacific Oceanic islands (Fig. 8). Tennessee is in the Nearctic biogeographical realm, which includes Canada, United States, and northern to central portions of Mexico.

All organisms contribute to total biomass (total weight of organisms in a defined space or volume) and either primary or secondary production in the habitats they occupy. Primary production is the synthesis of organic compounds from inorganic compounds. This conversion is accomplished by photoautotrophs and chemoautotrophs through photosynthesis and chemosynthesis. Photosynthetic organisms use sunlight as an energy source to convert the carbon of carbon dioxide into glucose. Chemosynthetic organisms obtain energy from the oxidation or reduction of certain inorganic chemical compounds, such as ammonia or hydrogen sulfide. Animal life is dependent on primary production.

Secondary production is the synthesis of heterotrophic biomass, which occurs when heterotrophic organisms consume autotrophs or other heterotrophic organisms. Secondary production involves a flow of energy between trophic levels of a food web.

How important are the contributions of freshwater sponges to biomass and secondary production in freshwaters? In most freshwater environments, sponges contribute a small amount to total benthic biomass (the total weight of the organisms living on the bottom of a body of water) and secondary production. However, that is not always the case. Sponges have been found to comprise a major portion of the invertebrate fauna biomass and contribute a significant amount to benthic secondary production in the Thames River of England.

Within the United States freshwater sponges have received little attention relative to other groups of aquatic invertebrates. This lack of attention has resulted in an absence of basic, but essential, information needed for the management and conservation of freshwater sponges. A report by the US Fish and Wildlife Service, in its mid-1970s review of freshwater sponges for listing purposes, recognized the limitations of existing knowledge. This report addressed the fact that conservationists have done little since an initial assessment to monitor or further evaluate the status of freshwater sponges within the

United States. One way to rectify this situation is to teach future generations of biologists the biology of freshwater sponges.

Freshwater sponges perform several functions within the ecosystems they occupy. These are:

1. As filter feeders, sponges actively and efficiently pump and clean large amounts of water.
2. Their pumping activity helps circulate the water surrounding them, especially in lentic waters.
3. Their pumping activity traps particulate and dissolved organic matter, thus they play an important role in the recycling of organic material.
4. They provide a living refuge for a wide array of organisms, such as viruses, bacteria, fungi, algae, protists, and larvae of caddisflies and spongillaflies.
5. They contribute to primary production by having symbiotic relationships with autotrophic microorganisms.
6. They contribute to the formation of sediments by releasing spicules upon their death.

Cherokeesia armata, Nolichucky River, Greene County, Tennessee.

Additional information for the conservation and management of freshwater sponges in Tennessee is needed. We have a list of species and limited distribution information. The earliest published papers concerning freshwater sponges in Tennessee are those of Clayton Hoff and J. G. Parchment. Hoff documented four species from the Reelfoot Lake region: *Eunapius fragilis* (Leidy, 1851), *Heteromeyenia tubisperma* (Potts, 1881), *Racekiela ryderi* (Potts, 1882), and *Radiospongilla crateriformis* (Potts, 1882). Parchment reported *Spongilla lacustris* from Stones River.

The first statewide survey of the freshwater sponges of Tennessee began in 2012. This survey resulted in the discovery of a new genus, *Cherokeesia* Copeland, Pronzato, and Manconi, 2015, and two new species, *Cherokeesia armata* Copeland, Pronzato, and Manconi, 2015, and *Heterorotula lucasi* Manconi and Copeland, 2022. *Cherokeesia armata* represents the first living member of the family Potamolepidae found in the Nearctic Region. The finding of *Heterorotula lucasi* enlarged the geographic range of this genus. *Heterorotula* was previously known from Australia, New Zealand, New Caldonia, and as an alien species in Japan.

Currently, 14 species of freshwater sponge are known to occur in Tennessee waters.

CHAPTER 2

Anatomy

"Take you me for a sponge."
—WILLIAM SHAKESPEARE—

When compared to other animals, sponges, both marine and freshwater, have some unique anatomical and physiological features. Several types of sponge cells are known to be totipotent. Totipotent cells are stem cells capable of changing morphology and physiology to become other types of cells. Totipotent cells occur throughout the animal kingdom but are uniquely different in sponges. The totipotent cells of other animals change in only one direction, from stem cell to the derived cell. In sponges these cellular changes can occur naturally in both directions. Interestingly, human stem cells can be manipulated after differentiation has occurred to convert back to the stem cell state, but this has not been observed to occur naturally.

An old but frequently asked question concerns the body of a sponge; is it a colony or an individual? The body bauplan (architecture) of a sponge does not precisely conform to either option. Instead, a sponge should be thought of as having a modular body design. Why? Because a sponge can alter its bauplan in time and space. Consider a carpet of gemmules formed by a mother sponge (Fig. 9). Each gemmule is a clone of the others and represents a ramet (a single member of a clonal group) capable of regenerating another sponge. As this carpet of gemmules germinate, the small individual sponges fuse to

FIG. 9. Basement membrane of gemmules.

create a genet (a fused clonal group of ramets) or single composite sponge. Each ramet of the genet represents a modular unit.

If you have played with wooden blocks, you have used modular units to build a block structure. Other modular organisms are Bryozoa (phylum of moss animals), some Cnidaria (phylum of animals possessing cnidocytes, or stinging cells, e.g., a jellyfish), and some Tunicata (subphylum that includes animals with dorsal nerve cord and notochord). A major difference between sponges and these three groups of animals is individual ramets of a sponge can be isolated from one another.

The body of a sponge is composed of three layers, pinacoderm, mesohyl, and choanoderm (Fig. 10). Pinacoderm is composed primarily of pinacocytes and porocytes. Pinacocytes are cells that attach a sponge to its substrate and line water canals. Porocytes are tubular cells that line pores found on the surface of a sponge and are involved with regulating water flow. The smaller pores are known as ostia (singular "ostium") and the larger as oscula (singular "osculum")

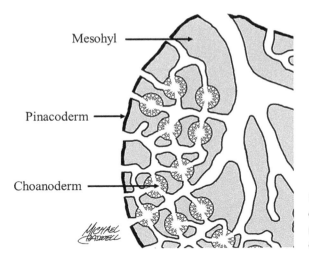

FIG. 10. Layers of a sponge body: pinacoderm, mesohyl, and choanoderm.

(Fig. 11). Ostia and oscula represent the openings at opposite ends of water canal systems. The water canal system is one of the defining features of a sponge. Ostia are intake openings through which water enters a sponge and oscula are outflow openings through which water exits the sponge. To visualize this, think of a water hose connected to a faucet and a lawn sprinkler. The connection at the faucet represents an ostium, the connection at the sprinkler represents an osculum and the hose between the connections represents the water canal.

Associated with porocytes are small muscle-like cells known as myocytes. Myocytes possess actin and myosin fibers, as do human muscle cells, which provide these cells with the ability to contract. Myocytes regulate waterflow through a sponge by regulating the size of ostia and oscula. Water flow into, within, and out of a sponge is vital for feeding, obtaining oxygen, releasing carbon dioxide and waste, sperm release and capture, release of larvae, and osmoregulation.

The mesohyl is a gelatinous matrix located between the pinacoderm and mesoderm. Several specialized cells, spongin, and skeletal components are found in the mesohyl. Amoebocytes are motile cells capable of moving in an amoeboid manner. Several forms of amoebocytes occur within a sponge, one of which is the archeocyte. These cells are totipotent cells capable of changing their morphology and function to become other types of cells. Cells derived

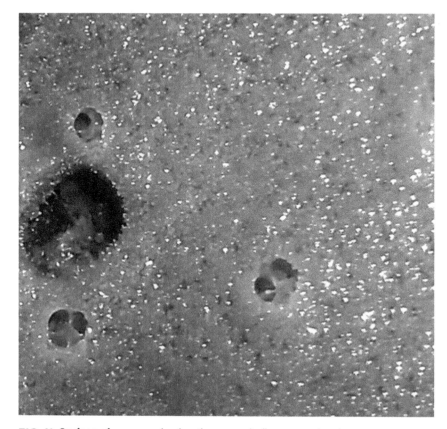

FIG. 11. Surface of a sponge having three oscula (large openings) and numerous ostia (pinprick openings).

from archeocytes are thesocytes, lophocytes, sclerocytes, and myocytes. Additionally, archeocytes are phagocytic which means they are capable of engulfing and processing food substances.

Freshwater sponges have a skeleton composed of spongin and siliceous spicules. Spongin is a modified form of collagen, a type of protein produced by collenocytes and lophocytes. It is elastic and exists in two forms, fibers and a thick vitreous gelatinous substance that forms the matrix of the mesohyl. Spongin fibers bind spicules together to form the skeletal frame. Additionally, spongin is important in the construction of gemmules. Spicules are produced

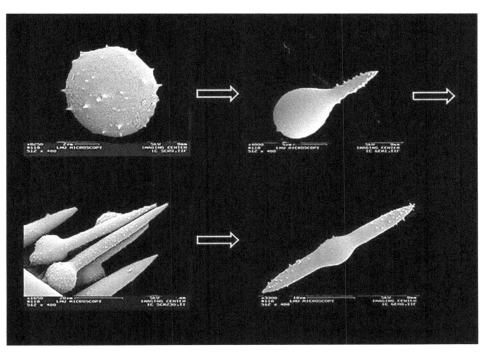

FIG. 12. Sequential formation of spicules by sclerocytes.

by cells known as sclerocytes. Spicules comprise the ridged, mineral portion of the skeleton. Spicules of a freshwater sponge form as sclerocytes secrete silica dioxide, SiO_2, along an axial organic filament (Fig. 12). Spicule production can be fast. *Ephydatia muelleri* can produce a spicule 200-350 µm in length and having a 15µm thickness in about 24 hours.

The rigidity or stiffness of the skeleton varies among species and is determined by the degree of interlocking or fusion of spicules and the availability of silica. Spicules support the mesohyl, provide protection to amoebocytes and thesocytes located within gemmules, and may act to prevent or reduce predation. The beauty of a freshwater sponge can be seen in its spicules. Some spicules are ornate in design.

There are three classes of spicules: megascleres, microscleres, and gemmuloscleres (Fig. 13). Megascleres form the primary skeletal support of a sponge.

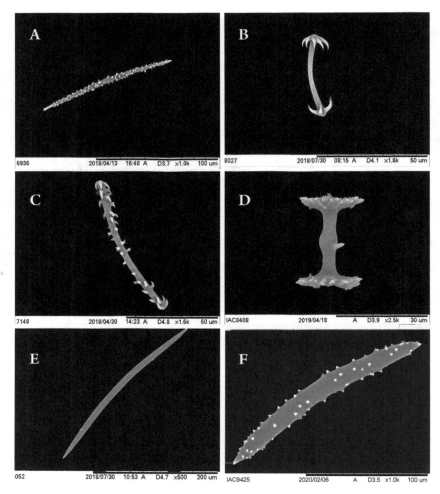

FIG. 13. Types of Spicules. **A-B**, microscleres; **C-D**, gemmuloscleres; **E-F**, megascleres.

Microscleres, which some sponges do not produce, provide secondary reinforcement. Gemmuloscleres form the protective armor-like coat of gemmules.

Because they are produced in a variety of sizes and forms (needlelike, spindle-like, spiny, smooth), spicules, especially gemmuloscleres, are important for sponge identification and taxonomy. Gemmulosclere shape is often species-specific. Spicules can be viewed using a light compound microscope (LM) but

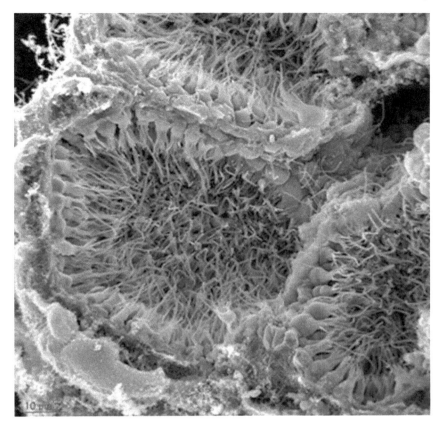

FIG. 14. Scanning Electron Microscope image of a choanocyte chamber of a leuconoid sponge.

because of their small size, measured in microns (μm), and the need to see details, they are best observed with a scanning electron microscope (SEM).

Choanoderm is composed of cells known as choanocytes or collar cells (Fig. 14). Choanocytes are multifunctional cells responsible for capturing, ingesting, and processing foods into nutrients, maintaining the flow of water through a sponge and transforming into male gametes. Choanocytes are flagellate cells having collars composed of microvilli (Fig. 15). The upper two-thirds of a microvilli collar is tightly held together by a glycocalyx mesh, while the lower one-third has relatively large openings between adjacent microvilli. Flagella are vaned (winged shaped). In *Spongilla lacustris*, collars average 8.2 ± 0.2 μm

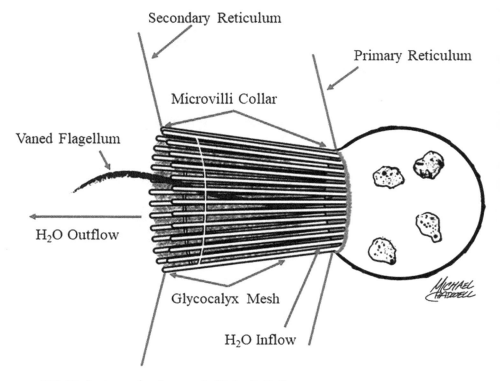

FIG. 15. Anatomy of a choanocyte (Collar Cell), the sponge pump.

in length, 3.1 ± 0.1μm in width at the collar base, flagellar central length 10.4 ± 0.3 μm and contains 24-36 microvilli. Frequency of flagellar beat in actively beating cells has been found to vary by a factor of seven, ranging from 3.2 to 20.9 hertz (Hz). One Hz is the equivalent of one event per second.

Based on the complexity of their water canal systems, sponges are divided into three types: asconoid, syconoid, and leuconoid.

Asconoid sponges have a canal system characterized by having pores extending directly from the surface of a sponge into a large cavity known as a spongocoel, which is lined with choanoderm (Fig. 16). This is the simplest type of canal system. Water enters through ostia, flows into the spongocoel, and exits through an osculum.

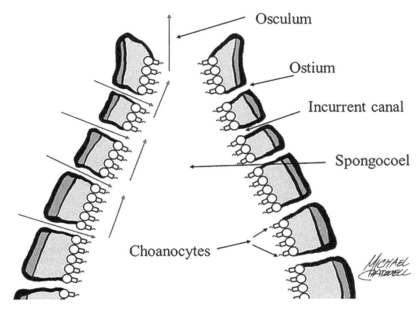

FIG. 16. Asconoid water canal system, arrows indicate flow of water.

The syconoid design is essentially a folded version of the asconoid type (Fig. 17). Syconoid sponges have a spongocoel, which is not lined with choanocytes, and two types of canals, incurrent and radial. Radial canals are lined with choanocytes. The opening of an incurrent canal leading into a radial canal of a syconoid sponge or choanocyte chamber of a leuconoid sponge is called a prosopyle. From radial canals and choanocyte chambers, water enters excurrent canals through small openings called apopyles. Water flow in a syconoid sponge is as follows: enters through ostium → incurrent canal → prosopyle → radial canal → apopyle → spongocoel → exits through osculum.

The leuconoid canal system is characterized by additional folding beyond that of the syconoid type. They have a greatly reduced spongocoel, branched canals, and choanocyte chambers. In leuconoid sponges excurrent canals unite to form larger canals that open into a spongocoel. Water flow in a leuconoid sponge is as follows: enters through ostium → subdermal space → incurrent canal → prosopyle → choanocyte chamber → apopyle → excurrent canal →

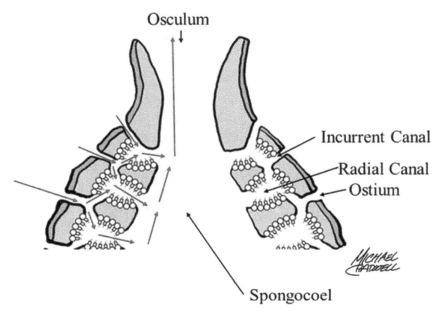

FIG. 17. Syconoid water canal system, arrows indicate flow of water.

exits through osculum (Fig. 18). Freshwater sponges have a leuconoid design. Due to volume constraints on the body of a sponge, the leuconoid body form is the only way to become large and complex. Increased size and complexity results from an increase in circulation which delivers more oxygen and nutrients.

The larvae of sponges of the genus *Spongilla* have a rhagon canal system (Fig. 19). The body is conical in shape with an osculum at its apex and a row of small choanocyte chambers in its walls. Water flow in a rhagon larva enters through ostia → subdermal space → prosopyle → choanocyte chamber → apopyle → spongocoel → exits through the osculum. The rhagon design is a developmental stage in the formation of a leuconoid water canal system.

From an evolutionary perspective, the question of which canal system gave rise to the others is yet to be resolved. There is evidence that points to the syconoid and leuconoid systems arising independently from the ascon type sponges and some evidence that supports leucon sponges giving rise to ascon and sycon sponges. Recent investigations support the view that sponge water systems evolved from the ascon type.

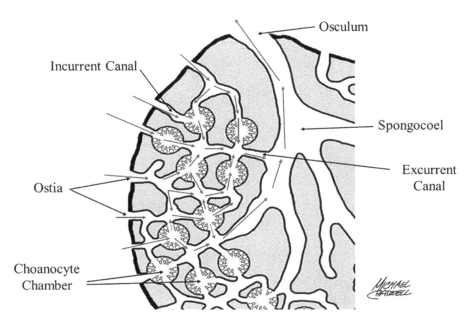

FIG. 18. Leuconoid water canal system, arrows indicate flow of water.

The study of water flow and fluid dynamics within choanocytes has primarily been restricted to theoretical calculations. It had been thought that the choanocyte chambers of a leuconoid sponge were the pumps controlling water flow. Additionally, it was believed that the choanocytes of a specific chamber acted as a synchronized unit. However, video recordings of choanocyte chambers reveal flagella beat asynchronously and at different rates. A model known as the "leaky pump model" suggests that the collar-flagellum component of each choanocyte is the actual pump unit as this is where the positive water displacement mechanism is thought to occur. Based on this model, collar pumps operate independently in parallel, with each delivering the pressure needed to propel water through a sponge. Collar pumps produce suction for water inflow and pressure for water to exit the sponge. Water arrives at a choanocyte chamber by way of incurrent canals and enters the chamber through prosopyles. Choanocyte chambers are designed to have sealed zones of low pressure and high pressure. Once in a chamber, water is sucked into each choanocyte collar through openings which exist between adjacent microvilli in the lower third

ANATOMY

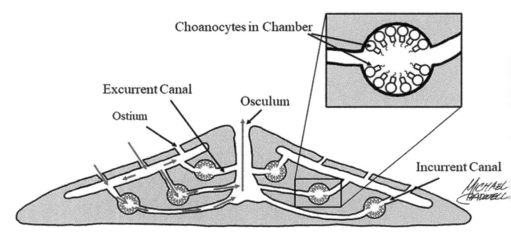

FIG. 19. Rhagonoid water canal system, arrows indicate water flow.

of the collar. Suction is generated by the action of the flagellum, which also causes the pressure in a collar to increase as water passes through the glycocalyx section. The function of the glycocalyx mesh is to seal the distal portion of the collar to prevent leakage while allowing water pressure to increase within the collar. This mesh also provides resistance from pressure forces within the collar, which could cause spreading of the microvilli. A low pressure is maintained within the collar between two structures located at opposite ends of the collar named the primary and secondary reticula. The primary reticulum is located at the base of the microvilli collar and the secondary reticulum is located at the distal ends of adjacent choanocyte collars. The secondary reticulum acts as a seal that separates the low-pressure zone in the collars from a zone of high pressure located between the secondary reticulum and the cone cell ring of each chamber. The high water pressure propels water through an apopyle, into the excurrent canals, and out of the sponge. This model hypothesizes that to generate high pressure, each choanocyte operates as a leaky, positive displacement-type pump. This pump operates by the interaction between the beating of the flagellum and the microvilli collar. The model gets its name from the backflow of water caused by small gaps between the flagellum and the collar. More work is required to fully understand the operation of the leuconoid water pump system.

Sponges have a remarkable ability to regenerate lost body sections. Even sponges forced through a sieve have reconstituted themselves! The actions of insects and other animals that eat or live on or within a sponge can create sponge fragments as can substrate tumbling, caused by high water flow and volume. Sponges are capable of regenerating not only lost or damaged sections, but also can transform a single cell or a sponge fragment into an adult sponge. How is this possible? As previously mentioned, sponge cells are totipotent. Their ability to change cellular morphology and physiology to become other types of cells gives them a major role in regeneration and development of a new sponge.

THREE

Natural History

"In all works on Natural History, we constantly find details of the marvelous adaptation of animals to their food, their habits, and the localities in which they are found."

—ALFRED RUSSEL WALLACE—

Natural history is concerned with the "what and how" questions regarding the actions of organisms. This includes life history information such as reproduction, dispersal, obtaining food, interaction with other life-forms and the environment, origins and evolution, and group organizations, such as populations and communities.

Life Cycle

The life cycle of a freshwater sponge alternates between times of active growth and dormancy. Three developmental stages (adult, larva, and gemmule) and four life phases (active and rapid growth, vegetative, degeneration and gemmulation, and quiescence) comprise this cycle (Fig. 20). The active and rapid growth phase represents the transformation of a sexually produced larvae, through metamorphosis, or asexually produced gemmule, by germination, into a sponge. The vegetative phase is the growth phase of a sponge. Depending on the availability of food and size of substrate a sponge can grow to a large size in just a few months. Usually during late summer and fall environmental conditions become unfavorable for the survival of adult sponges resulting in degeneration and gemmulation. The degeneration and gemmulation phase

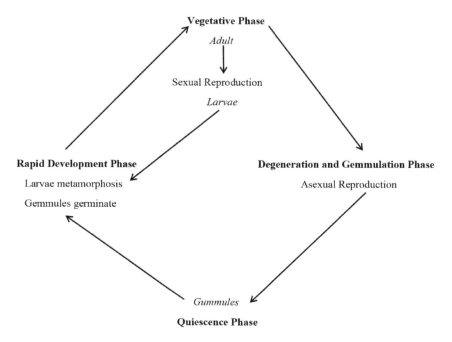

FIG. 20. Annual life cycle of freshwater sponge. Phases in bold print and stages in italics.

involves the dissolution of the adult while at the same time, the production of gemmules. Initially, cells found within a gemmule are in a state of suspended metabolic activity, a condition known as diapause. Diapause is controlled by endogenous factors. After exposure to unfavorable environmental conditions cells enter a second phase of dormancy known as quiescence. Quiescence differs from diapause in that it is controlled by exogenous (environmental) factors. Gemmules remain in the quiescence phase until environmental conditions, such as warming water temperatures, become favorable for germination. At the end of the quiescence period gemmules germinate and the cycle repeats itself.

Endogenous processes, those occurring within a sponge and controlled by a biological clock, play a role in controlling phases of the life cycle. On the other hand, differences in seasonality and length of phases are influenced by exogenous limiting factors, such as ice-up and dry-up, which lead to hibernation or aestivation.

Reproduction

Sexually, freshwater sponges may be hermaphroditic or gonochorous. Hermaphroditism is the condition of having both male and female reproductive organs in a single individual. Gonochorism, also known as unisexualism, is the state in which an individual sponge is entirely male or female. Sex reversal has been observed in the freshwater sponge *Spongilla lacustris*. During one reproductive event this sponge produces sperm and during the next event, eggs. Organisms capable of sex reversal from one reproductive event to the next are known as sequential hermaphrodites. Gamete production occurs through the transformation of specialized cells, sperm from choanocytes and oocytes from archeocytes. Sexual reproduction is synchronized with eggs and sperm maturing at the same time, even though production of oocytes begins before that of sperm. Regardless of the sexual make up of an individual sponge, sexual reproduction occurs through cross-fertilization.

Adult sponges produce the next generation of sponges by sexual or asexual reproduction. Sexual reproduction results in progeny genetically different from their parents, a difference known as genetic diversity. Genetic diversity is necessary for the long-term survival of populations by providing them the genetic flexibility to adapt to and survive in changing environments. Produced through meiosis, sperm, and eggs each have a haploid (n) number of chromosomes (single set of chromosomes). Water currents transport sperm to the ostia of other sponges where choanocytes capture them. Choanocytes transport sperm to eggs located in the mesohyl where fertilization occurs, resulting in a zygote. Zygotes have the diploid (2n) number of chromosomes (double set of chromosomes) as they received a single set of chromosomes from each parent. Zygotes develop into parenchymella larva, which have a large vacuole, and a single cell layer of ciliated epithelium (Fig. 21). Freshwater sponges are viviparous as the mother sponge gives birth to live young. Once released from the mother sponge, larvae disperse by swimming and by water currents. Larvae represent the shortest stage of the life cycle, living for only hours to a few days. Larvae must attach to an appropriate hard substrate for metamorphosis to occur.

Asexual reproduction involves a single parent and in freshwater sponges can occur through three processes: gemmulation, fragmentation, and budding. Asexual reproduction is nature's way of cloning and does not result in genetic diversity.

In temperate climates gemmules (Fig. 22) are usually produced during summer and fall in preparation for the quiescence phase of the life cycle. Gemmule

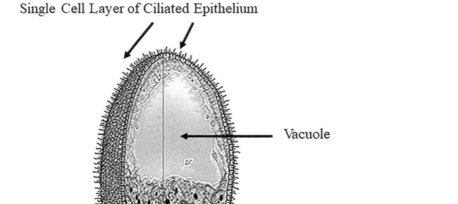

FIG. 21. Illustration of Parenchymella larva.

production can be rapid and abundant. A specimen of *Ephydatia fluviatilis* only five centimeters in diameter may produce more than 2,000 gemmules in less than a week.

Gemmules have two functional roles: resting bodies and propagules. As resting bodies, gemmules are survival structures that allow sponges to survive the adverse environmental conditions of freezing, desiccation, and anoxia, which are detrimental to the adult sponge. Gemmules are formed as the sponge body begins to disintegrate. If you are a fan of science fiction, think of a gemmule as an escape pod released from a dying planet (Fig. 23). The timing of the quiescence phase has been found to vary for a specific species occupying different habitats. *Ephydatia fluviatilis* is an extremely adaptive and widespread species. In latitudes above 40° where harsh winters are encountered, *E. fluviatilis* produce gemmules in preparation for hibernation. On the other hand, *E. fluviatilis* occurring in environments of high temperatures and desiccation

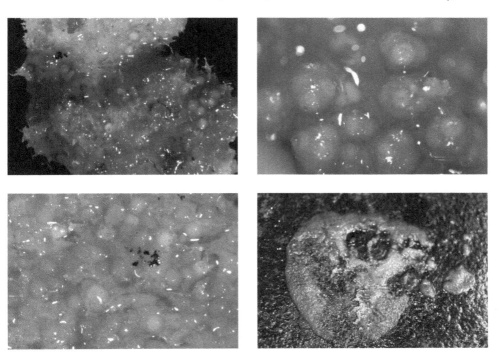

FIG. 22. Gemmules in vivo.

produce gemmules in preparation for estivation. Additionally, the timing of dormancy has been found to vary among species found in the same body of water. In Lake Pontchartrain, Louisiana, *Spongilla alba* Carter, 1849 is dormant during winter months while *E. fluviatilis* exhibits summer dormancy.

As propagules, gemmules are reproductive structures that contain thesocytes surrounded by a protective coating of interlocking gemmuloscleres located between two spongin layers. Thesocytes are derived from archeocytes and are structurally similar except that they lack phagosomes. Phagosomes are vacuoles, which are capable of ingesting food particles. Thesocytes do not need phagosomes because they contain stored food in the form of vitelline platelets.

Gemmule formation begins when archeocytes and nurse cells known as trophocytes aggregate together creating a mass. Archeocytes obtain polyols (sugar alcohols) from trophocytes. Polyols are used to form vitelline platelets (yoke producing cells). Archeocytes filled with vitelline platelets transform

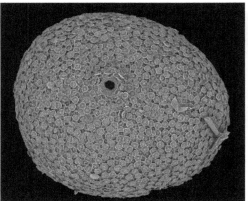

FIG. 23. Gemmule compared to a space capsule.

becoming binucleated thesocytes. Amoebocytes surround the mass of developing thesocytes and secrete an inner and outer membrane composed of spongin. Sclerocytes secrete gemmuloscleres between the two membranes (Fig. 24). Fully developed gemmules are spherical to ovoid in shape and have an opening called the foramen.

Once gemmules are formed the thesocytes within them enter diapause, a state of suspended development controlled by endogenous factors. One factor appears to be osmotic pressure. Thesocytes newly derived from archeocytes have a high osmotic pressure because they contain a high concentration of polyols. The high osmotic pressure prevents cell division and causes gemmules to remain in diapause. The quiescence phase is reached by thesocytes lowering their osmotic pressure, which happens by converting polyols to glycogen. This conversion also restores cell division. Quiescence is maintained until water temperatures begin to rise, which promotes germination. In *E. fluviatilis*, as germination begins, binucleated thesocytes undergo cytokinesis to become uninucleated archaeocytes which are released from the gemmule. After release the archaeocytes undergo proliferation and differentiation to begin the development of a new sponge. A single gemmule has the potential to give rise to several young sponges.

Gemmules produced by sponges living in aquatic environments where a complete dry-up or very low water levels occur, break quiescence with the return of water. Other species of freshwater sponges cannot survive desiccation.

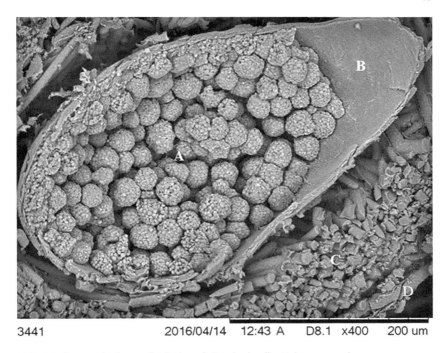

FIG. 24. Gemmule theca. **A**, cluster of staminal cells; **B**, inner membrane of spongin; **C**, gemmuloscleres; **D**, outer membrane of spongin.

Such sponges have their ranges restricted to areas such as deep lakes, where there is little chance of drying up.

Fragmentation occurs when a portion of a sponge becomes separated from the rest of the body. Body fragments can be dislodged by actions of insect larvae and other animals that feed on or live within sponges, and through abrasion, such as rock tumbling caused by high water velocity. Fragments that settle on a suitable substrate can develop into separate sponges.

Budding is the development of a new individual from an outgrowth which develops on an adult sponge. Budding has been observed in *Radiospongilla cerebellata* (Bowerbank, 1863). In *R. cerebellata* bud formation begins as a raised area on the sponge surface. This area develops rapidly forming a broad base that grows to extend above the surface of the sponge. As the base is elevating above the sponge, a bud develops, and the base narrows to become a stalk that breaks to release the bud. Buds are carried by water currents to suitable substrates where they transform. Bud size for *Radiospongilla cerebellata* ranges from 1.5 to 2.5 millimeters in diameter.

Dispersal

Biological dispersal refers to the movement of individuals or gametes away from their place of origin. Range expansion occurs when individuals disperse to new unoccupied areas. The ability to disperse is an important factor in determining the distribution of a species.

Dispersal is vital for long-term survival of populations by facilitating gene flow between them. Dispersing individuals may carry genes not found in the population receiving them. If these individuals become involved in reproduc-

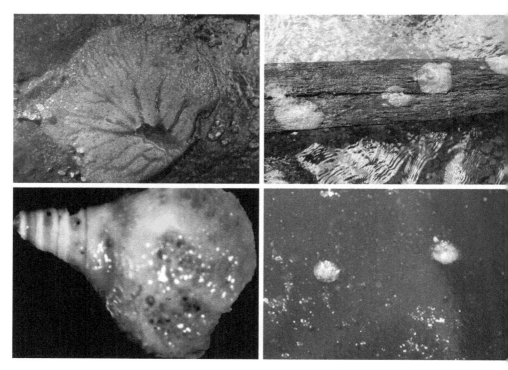

FIG. 25. Freshwater sponges attached to hard substrates. **A**, rock; **B**, wood; **C**, shell of aquatic snail; **D**, two small white sponges attached to hard plastic.

FIG. 26. Caddisfly larva, *Ceraclea sp.*, with case composed of sponge.

tion, they could potentially introduce new genes into the population. Also, as habitats change over time, by natural or human influences, they may become so altered a population can no longer be supported. Individuals dispersing to new areas reduce the chance of extinction.

The availability of hard substrate is essential for sponge dispersal (Fig. 25). Sponges must attach to a suitable substrate before individual development and population growth can occur. For many freshwater sponges, a sparseness of hard substrate places limits on population growth. Only two sponges, *Spongilla lacustris* and *Ephydatia muelleri*, have been documented to colonize habitats lacking rocks, logs, or other hard objects.

Dispersal typically occurs by water currents transporting sponge fragments, larvae, and gemmules downstream to new substrates. However, actions of larvae of sponge eating caddisflies of the genus *Ceraclea* can result in sponge dispersal. Some species of *Ceraclea* use sponge materials in the construction of their larvae cases (Fig. 26). They have the remarkable ability to interconnect spicules in an orderly fashion to create an ornate case. Before pupation, mature larvae usually leave their sponge host transporting sponge material with

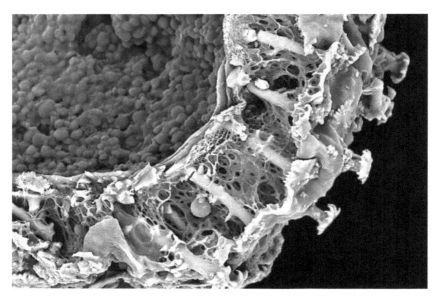

FIG. 27. Cross section of a gemmule having a pneumatic membrane (in light puce color) embedded with gemmuloscleres (in tan to green color), inner and outer spongin layers (in light blue color) and thesocytes (in dark blue color).

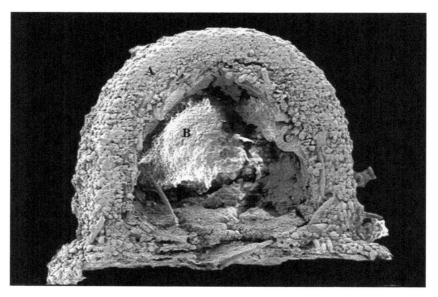

FIG. 28. Cross section of gemmule lacking a pneumatic membrane. **A**, gemmule wall composed of gemmuloscleres; **B**, staminal cells; **C**, inner membrane.

FIG. 29. Surface of gemmule; spongin (brown) and gemmuloscleres (pink).

them. Larvae find suitable substrates and undergo pupation. After maturing, the adult caddisfly leaves the pupal case, which remains attached to the substrate. Sponge tissue may spread from the pupal case to the substrate, thereby resulting in dispersal.

How do freshwater sponges disperse upstream, against the water current? While no evidence currently exists, birds and mammals may be responsible. It is thought gemmules could become attached to feet, legs, and feathers of ducks and wading birds, and hair of mammals, and be transported not only upstream, but also to other waterways. Are some gemmules capable of passing intact through an animal's digestive tract? If so, gemmules could be transported upstream by sponge-eating animals. Wind may be involved in upstream dispersal. During droughts or low water levels, sponges become exposed to the atmosphere as water levels drop below their substrates. Light weight gemmules could be lifted from a substrate by high winds and transported elsewhere.

Gemmule dispersion is facilitated by the presence of a pneumatic layer within the gemmular theca. (Fig. 27, Fig. 28, and Fig. 29). Members of the family Potamolepidae produce gemmules lacking a pneumatic layer, or one scantly developed, while most members of Spongillidae produce pneumatic gemmules. Pneumatic gemmules are designed for overland dispersal. Overland dispersal occurs by water currents transporting gemmules downstream. On the other hand, gemmules lacking a pneumatic layer are not designed for overland dispersal. They disperse when their substrates are moved by a physical force such as highwater velocity or when they are dislodged when struck by some force or object.

Foods and Food Webs

Sponges are filter feeders. Particles as small as one micron to those too large to be taken into the sponge through ostia can be consumed. Foods consumed include bacteria, phytoplankton, and detrital organic particles. Sponges are selective feeders. Studies have shown sponges can selectively remove particles present in small concentrations while food particles present in larger concentrations pass through the sponge.

Foods are removed from water as it passes through choanocyte chambers. In addition, porocytes found on the sponge surface and pinacocytes lining canals can capture food particles through phagocytosis.

Because food particles are filtered from highly diluted suspensions, sponges must filter large volumes of water. Sponges filter a water volume six times, or more, their body volume per minute. *Spongilla lacustris* can filter water at a rate of more than 6 ml/hour per milligram dry mass of tissue. To put this into context a finger size sponge has the potential to filter more than 125 liters (33 gallons) of water in a 24-hour period.

Water and food inflow result from suction through ostia. Once inside a sponge, water passes through a series of progressively smaller canals until arriving at a choanocyte chamber. Choanocytes capture and filter foods using their microvilli collars. Microvilli secrete mucus, which traps particles as filtration occurs through gaps between adjacent microvilli. Captured food particles are consumed by choanocytes through phagocytosis. Once digested, nutrients are passed to archeocytes that move about within the mesohyl transferring nutrients to other sponge cells. Non-digested food products are removed from the sponge by archeocytes releasing waste products into excurrent canals that connect to oscula, from which they are expelled from the sponge.

Biologists construct food webs to show the interconnected feeding relationships and flow of energy in an ecosystem. Food webs illustrate the transfer of organic food energy from its source in primary producers (plants and other autotrophs) to heterotrophic primary consumers (herbivores) to secondary, tertiary, and quaternary consumers (carnivores) and to decomposers (bacteria and fungi).

Freshwater sponges are heterotrophic consumers. In addition, those having algae endosymbionts illustrate a mixotrophic mode of nutrition. Mixotrophic refers to sponges receiving nutrients from two sources, algae symbionts living within them, as well as through their own pelagic food sources.

Freshwater sponges serve as a link among benthic, pelagic, and terrestrial food webs. This is a result of a tri-trophic relationship involving freshwater sponges, phytoplankton, and sponge eating animals. Freshwater sponges are known to feed largely on pelagic food sources, such as phytoplankton. Sponges can be consumed by benthic organisms such as crayfish, thereby transferring energy obtained from phytoplankton to the crayfish linking pelagic and benthic food webs. A crayfish in turn may be consumed by a raccoon, *Procyon lotor* Linnaeus, 1758, a terrestrial mammal.

Aquatic larvae of some groups of insects feed on freshwater sponges. The larvae of spongillaflies and some caddisflies can have a major portion of their diets consisting of sponges having alga symbionts. Thereby, benthic sponges, and their algal endosymbiont provide nutrients to insect larvae, which in their adult forms may be consumed by terrestrial animals such as birds and spiders.

Interactions with Other Life-Forms

As previously mentioned, freshwater sponges often have symbiotic associations with green algae, thus it is not uncommon to find a green colored sponge (Fig. 30). Sponges and algae share a long evolutionary history. Unicellular green algae of the genus *Chlorella* form intracellular symbiotic associations with several species of freshwater sponges. A yellow-green alga is associated with *Corvomeyenia everetti* (Mills, 1884). The evolutionary history of *Corvomeyenia* with its yellow-green alga symbiont is distinct from that of freshwater sponges having green-algae symbionts. While these two evolutionary histories are similar, they represent convergent symbiotic associations that have evolved separately. The symbiotic association between sponges and algae is an example of facultative

FIG. 30. Freshwater sponges having green algal symbionts.

mutualism, which is a biological relationship in which both organisms benefit from the association. However, the relationship is not essential as each organism can live and survive independently of the other.

A benefit of this association, as stated previously, is mixotrophic nutrition. Mixotrophic nutrition is beneficial to both organisms. Algae receive sponge produced carbon dioxide, nitrogen, and phosphorus and sponges receive photosynthetically produced glucose. A study of *Ephydatia fluviatilis* and its green alga symbiont found 9% to 17% of the total glucose produced by algae was provided to the sponge. Thus, sponges utilize organic energy compounds produced autotrophically and heterotrophically. Mixotrophic nutritional associations have proven beneficial in nutrient-poor habitats.

The degree to which sponges rely on algal symbionts for nutrients ranges from none to a substantial amount. Researchers studying *Spongilla lacustris* in a New Hampshire Pond determined that algae produced carbohydrates accounted for 50% to 80% of the growth of the population of this sponge. Exposure to sunlight, regardless of its intensity, causes an increase in algae production, which results in higher growth rates for sponges. On the other hand, sponges living in complete darkness rely exclusively on animal heterotrophy.

Diverse communities of bacteria are known to occur on and in sponges. Studies of microbial community structure in freshwater sponges using Next Generation Sequencing (NGS) technology has revealed just how diverse these communities can be. Using this technology, DNA of bacteria belonging to 14 phyla have been found associated with *Eunapius carteri* (Bowerbank, 1863) and *Corvospongilla lapidosa* (Annandale, 1908). Phyla discovered included Actinobacteria and Cyanobacteria. This is an important finding because it could lead to the discovery of compounds having biotechnological uses. Actinomycetes, a group of bacteria of the phylum Actinobacteria, have had a major role in the development of antibiotics. The actinomycetes are characterized by being filamentous or rod-and-coccus in appearance, having lateral protuberances, and being gram-positive. About 45% of all the bioactive compounds that have been produced from microbes have come from actinomycetes.

The finding of Cyanobacteria is interesting as some members of this phylum are capable of nitrogen fixation. Nitrogen fixation is the process in which atmospheric nitrogen (N_2) is converted into ammonia (NH_3). Because atmospheric nitrogen does not easily react with other chemicals to form new compounds, it must be placed in a compound, such as ammonia, which is more reactive.

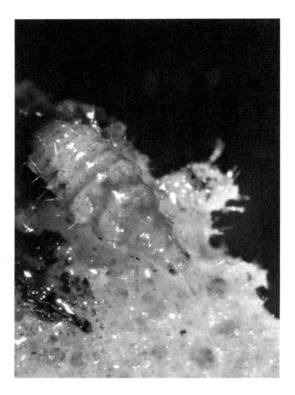

FIG. 31. Spongillafly larva.

The process of nitrogen fixation is vital to life on earth because nitrogen is essential for living organisms to produce amino acids, nucleic acids (DNA and RNA), and proteins. There is some limited evidence of nitrogen fixation by cyanobacterial symbionts in marine sponges. *Oscillatoria* is a genus that has members capable of fixing nitrogen. Some species of *Oscillatoria* are symbionts of the microbiomes of sponges. Sponges having cyanobacteria symbionts are known as cyanosponges.

A large and diverse group of aquatic insects have associations with freshwater sponges. Species of Coleoptera (beetles), Diptera (two-winged or true flies), Ephemeroptera (mayflies), Hemiptera (true bugs), Lepidoptera (aquatic moths), Megaloptera (alderflies, fishflies, Dobsonflies), Neuroptera (spongillaflies) (Fig. 31), Odonata (dragonflies), Plecoptera (stoneflies), Trichoptera (caddisflies), and Trombidiformes (aquatic mites) are associated with freshwater sponges.

FIG. 32. Zebra mussels, Dreissena polymorpha, notice variation in shell markings.

Attack of the Killer Sponge

The introduction of the non-indigenous zebra mussel, *Dreissena polymorpha* (Pallas, 1772), in the United States has had a negative impact on native unionid mussels (Fig. 32). Biologists have documented the decline and total disappearance of eleven unionid mussel species in Lake St. Clair, Pounce, Ontario after the arrival of the zebra mussel.

Zebra mussels are native to large rivers and lakes draining into the Azov, Black, and Caspian Seas of southwestern Russia and Ukraine. Zebra mussels are thought to have entered North American waters in the mid 1980s. Freshwater mussels, unlike marine mollusks, have not developed anti-fouling defense mechanisms. Unionid mussels encrusted with zebra mussels face a reduction in biochemical fitness and death. Biofouling mechanisms caused by zebra mussels growing on unionids are a hindrance to movement and burrowing, disruption of balance, invasive growth within the shells of unionids, siphonal interference, interference of feeding, deformation of unionid valves, and suffocation because of valve occlusion.

Freshwater sponges and zebra mussels are competitors for substrate space. Sponges often overgrow zebra mussels and encapsulate them, which can be lethal. Death occurs when the inhalant siphons of zebra mussels become occluded with sponge growth. An occluded inhalant siphon hinders respiration and feeding. Zebra mussels surviving epifaunal sponge growth often have reduced fitness when compared to non-sponge associated mussels. Loss of fitness is due to a reduction in glycogen content and development of soft tissues. A study within the Great Lakes–St. Lawrence River system found mussel colonies overgrown for one or more months contained a significantly greater proportion of dead mussels than uncovered ones, and mussels surviving four to six months of overgrowth suffered significant tissue weight loss. Freshwater sponges have the additional advantage of surviving desiccation. Although freshwater sponges have the competitive advantage, zebra mussels get a reprieve during the dormancy period of the sponge life cycle. A factor favoring zebra mussels is siltation. Sponges are intolerant of siltation. Siltation restricts sponge competition with mussels to vertical and other surfaces not covered by silt. It is believed freshwater sponges are unlikely to significantly impact zebra mussel populations in most lakes or rivers but may control zebra mussels in local habitats.

Sponges as Food and Habitat

Sponges provide a suitable microhabitat for a diverse assemblage of organisms such as endosymbiotic algae, viruses, bacteria, fungi, sponge predators, and organisms using the sponge as a selective microhabitat refuge. A study of the freshwater sponge *Corvospongilla siamensis* Manconi & Ruengsawang, 2012, in the Pong River of Thailand found it has associations with aquatic insects of 4 orders, 10 families and 19 taxa, some of which are predators. The larvae of spongillaflies (Neuroptera) (Fig. 31), some species of chironomid midges (Diptera) (Fig. 33), and caddisflies (Trichoptera) (Fig. 26) have mouth parts designed for feeding on sponges.

The caddisfly, *Ceraclea transversa* (Hagen 1861), which occurs in Tennessee, has a unique life history in that it has two distinct larval cohorts. The larvae

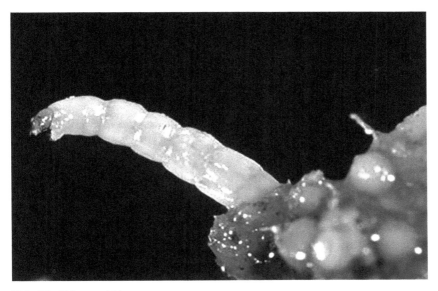

FIG. 33. Chironomid larva.

of the first cohort feed exclusively on freshwater sponges and overwinter as inactive prepupae, in the spring they pupate and emerge. The larvae of the second cohort feed on sponges until the onset of sponge gemmulation in the autumn at which time they leave the sponge. These larvae overwinter as active third-or-fourth instar detritus feeders and do not pupate and emerge until the following summer. Caddisflies digest sponge material, endocellular algae, and have the remarkable ability to ingest siliceous spicules. Their guts have a thick peritrophic membrane which prevents spicules from protruding through the gut wall.

Spongillaflies deposit eggs on vegetation overhanging a stream or lake and covers them with a web. When the larvae hatch, they drop into the water and search for a sponge. The larvae live within pores and natural cavities of sponges. Larvae have mouths designed to pierce and suck fluids from their host.

A limited number of animals other than insects have been reported to prey on freshwater sponges. Crayfish of the genus *Orconectes* consumed *Eunapius fragilis* and *Spongilla lacustris* in a Massachusetts stream. In Cameroon a fish, *Pungu maclareni* (Trewavas 1962), eats freshwater sponges. Freshwater sponges are one of the most common foods consumed by yellow-blotched map turtles, *Graptemys flavimaculata* Cagle, 1965, in Mississippi. One duck, the ring-neck duck, *Aythya collaris* (Donovan, 1809) is known to feed on freshwater sponges.

Unionicola crassipes (Muller, 1776) is a water mite that has a unique obligatory association with the freshwater sponge *Eunapius fragilis*. A two-year cycle from egg to adult for females of this mite has been documented in a pond in Alberta, Canada. Fertilized female mites overwinter and oviposit (lay eggs) in *E. fragilis*. Mite larvae emerge from sponges in May to early June to parasitize larvae of the chironomid genus *Tanytarsus*. After feeding, they leave the chironomid and search for a sponge. Upon returning to a sponge, mite larvae enter the protonymph stage. In early summer, protonymphs undergo transformation to become deutonymphs. Mite larvae overwinter in a sponge as deutonymphs. In early May of the following year deutonymphs transform to become tritonymphs. Adult mites emerge from the tritonymph stage in late May to early June. In September, fertile adult females return to overwinter and oviposit in a sponge, starting the cycle over again.

Some vertebrates use freshwater sponges for nesting and cover. In India, the fish *Gobius alcockii* Annandale, 1906, has been reported using *Eunapius carteri* for egg deposition and nesting habitat.

Sponge Distributions

Distribution refers to the geographical area in which individuals of a species are located. North American sponge species show a great deal of variation in their distribution patterns. In the United States some species such as *Ephydatia fluviatilis* are broadly distributed while *Corvomeyenia carolinensis* Harrison, 1971, is only known from a couple of localities. In Tennessee *Trochospongilla horrida* (Weltner, 1893), and *Eunapius fragilis* are widely distributed while *Corvospongilla becki* Poirrier, 1978, is known from a single location.

The distributions and abundance of organisms are determined by habitat suitability, their ability to disperse, and to cross geographical barriers. Habitat suitability is determined by biotic and abiotic factors. Biotic factors influencing distributions of freshwater sponges include competition between species and individuals for food and substrate, presence or absence of aquatic vegetation, predators, symbiotic algae, and diseases. Abiotic factors include climatic conditions, availability of nutrients, presence of silica and other minerals, salinity, pH, water temperature, rainfall, geological formations, silt accumulation, and available substrate. Investigations are needed to gain a better understanding of how factors act individually or in combinations to determine distributions of freshwater sponges.

As previously mentioned, dispersal is important in determining distributions. Landscape features such as isolated lakes and streams cut off from other bodies of water tend to support a less diverse sponge community or no sponges. Dispersal of gemmules or sponge fragments by wind or animal movements may be responsible for dispersing sponges upstream or to disjunct bodies of water. Downstream dispersal by water currents to interconnected streams and rivers is the more typical mode of dispersal.

Physical features of streams and rivers can influence sponge presence. For example, many streams of the Great Smoky Mountains National Park are characterized to be high gradient, thoroughly scoured, and nutrient-poor with limited ecological variability. Streams having such characteristics are not likely to support sponges.

Other known physical features influencing sponge distributions are availability of substrate, water level, silt, and sediment composition.

The availability of hard substrates is extremely important in sponge population dynamics and sponge community development. The life span of an individual sponge, including the transitions between active and dormant stages, is related to the permanence of its substrate. As noted previously, the availability of hard substrates is critical for the settlement, attachment, and growth of sponge larvae and gemmules. Natural objects such as rocks, boulders, logs, roots of living aquatic plants, and shell remains of living and dead aquatic mollusks are utilized as substrates. Human-built structures such as bridge foundations, walls of dams, piers, docks, and riprap provide suitable attachment surfaces. In southern Lake Michigan, sponge cover on revetment walls, stone riprap, and wooden pier posts ranged up to 13% of the available surface area, whereas no sponges were found growing on the soft silt bottom.

In rivers and streams, rocky shoals provide excellent freshwater sponge habitat, as hard substrate is abundant and water flow transports oxygen and food and flushes sponge waste, silt, and sediments downstream. Competition for space in habitats with limited substrate can influence not only sponge distributions but also population dynamics.

A study in Utah found sediment matrix composition, water depth, and silt deposition to influence sponge distribution. This study found the percentage of coverage of sponges to be three to thirty times greater on rocks located on hardpan than rocks located in gravel or sand. Water deeper than 1.2 meters had two to sixteen times greater sponge coverage than water less than 1.2 meters. Silt abundance increased gradually from the surface to a depth of 2.8 meters, after which there was a thick layer of ooze and silt, in which no sponges were found. Silt and sediment deposition can be harmful to sponges by occlusion of the sponge water canal system, complete smothering of the sponge body, and prevent larvae and gemmule attachment to substrata. Additionally, silt may cause sponges to colonize the vertical and bottom surfaces of rocks.

Water temperature influences the seasonality of *Ephydatia fluviatilis* and *Spongilla alba* differently. One study found *Spongilla alba* to not grow at 25° C, to grow slowly at 28° C, but then realize a 400% increase in growth as water reached 30° C, while *Ephydatia fluviatilis* grew in waters a cool as 16° C but did not grow at temperatures equal to or greater than 25° C.

The availability of silica can influence freshwater sponge distribution and development. Some species of freshwater sponges are adapted to occupying habitats high in silica and others to low levels, thus, silica concentrations may be a factor in the disjunct distribution of some species. Sponges, such as *Spongilla lacustris*, that have adapted to surviving in habitats of varying silica concentration may have a more continuous or wide distribution across landscapes but may have modifications in body consistency. In silica-poor habitats, sponges adapted to a wide range of silica produce less ridged, thinner, and more easily broken spicules. Thinning and breakage of spicules influences body consistency and possibly predator defense. Chapter 7 provides specific information concerning ranges of chemical and physical properties of water within which each of the fourteen species occurring in Tennessee are known to survive and reproduce.

Food availability is an obvious habitat component. Food quantity and quality affect growth, reproduction, and population size. Associations with intracellular green algae are beneficial to the growth of individual sponges, production of progeny, and population size. Without symbiotic associations with algae, sponges living in habitats low in food availability would show less growth, lower reproduction, and reduced population size.

Table B1 and B2 of the appendix provide distribution data for freshwater sponges found in the United States. Table B1 provides a distribution of freshwater sponge species by state. Table B2 provides a list of sponges for each state.

Mobility

Adult sponges are sessile. However, research has revealed that sponges are capable of movement. Sponge trails composed of spicules have been observed on seabeds in the Arctic Ocean.

Sponges lack muscles and nerves which coordinate movement in humans and other vertebrates. Without muscles and nerves how is it possible for a sponge to move? They move by sponge deflation.

Deflation results from tension release of actomyosin stress fibers in pinacocytes lining water canals. Molecular signals, like those involved with the inflammation process in humans, cause the actomyosin fibers of pinacocytes lining incurrent water canals to relax as those of the excurrent canals expand. Relaxation causes incurrent canals to collapse, which results in the sponge deflating. As incurrent canals collapse water is forced out through excurrent canals. This expulsion of water is the physical force which causes movement of the sponge. An easy way to picture this is to fill a balloon with air then release the ballon. Sponges are not "speed demons," moving less than 5 millimeters per day.

Contribution to Sediments

The body of a sponge disintegrates as the sponge prepares for hibernation or estivation. In this process spicules are released and settle onto the bottom of the body of water. Because spicules are composed of silica, they do not decay but become integrated into the developing sediments. Studies of spicules in paleo-sediments have proven to be a good diagnostic tool for determining the conditions of ancient lakes and marshes.

CHAPTER 4

Field and Laboratory Techniques

"Learning is the only thing the mind never exhausts,
never fears and never regrets."

—LEONARDO DA VINCI—

It is important to see and study organisms in their natural environment. Some forethought is essential and wise before heading into the field. It is not advisable to take solo field trips. Even if you are familiar with the area, hazards exist. A fall, snakebite, or other mishap can put one in a critical situation. A slip in a rocky creek can result in a serious head injury resulting in unconsciousness and possibly death by drowning. Even with cell phones, help can be minutes or hours away. It is wise to have at least one other individual accompany you and to inform others as to where you are going.

Give close attention to weather reports. Weather apps for cell phones are great for keeping track of upcoming weather. Doing so will provide you with time to leave the field before getting caught in a dangerous weather event. Those of us working with aquatic organisms know how quickly water levels in creeks and rivers can rise because of heavy rainfall. You do not want to be on the wrong side of a river before you realize the water has risen to a dangerous level and velocity.

Dress appropriately. When in the field expect to get dirty and wet. Footwear is especially critical for remaining upright when wading creeks. Select footwear that you find comfortable, provides support, and has good drainage.

Some people prefer felt soles as they provide a better grip on gravel and rock. However, on sand and mud substrata felt soles become slippery.

We always carry USGS topographical maps. Before the advent of handheld Global Positioning System (GPS) devices, a good compass and topo map allowed us to keep track of our location. However, the usefulness of a GPS device cannot be overstated. It is advisable to program a GPS unit to track your excursion and to lead you back to your starting point in case you get lost.

In Tennessee it is necessary to obtain a scientific collector's permit before collecting freshwater sponges or any other form of wildlife. In Tennessee to obtain a collector's permit, contact the Tennessee Wildlife Resources Agency.

Field Techniques

It is important that accurate notes be taken in the field while the work is underway or immediately after the work is completed. Waiting until later often results in inaccurate or incomplete information.

Your field notes should provide answers to questions concerning who, what, where, when, and how. "Who" information gives names of persons involved and their roles in data collection. "What" data includes the number of specimens collected, types of substrates collected from, types of habitats searched (pools, shoals, etc.), and the amount of time searching a specific site. "Where" information provides appropriate geographical information such as latitude and longitude, name of the body of water from which the sample was collected, name of drainage system (if collected from a stream or river), any landmarks specific to the collection site (such as a bridge name, road number, etc.), and county and state. "Where" information should be as precise as possible so that others can find your collection sites. "When" data includes collection date, time of day, and weather conditions. "How" information describes your search protocol (wading, scuba diving, etc.) for finding specimens.

Record your thoughts and observations. Do you have any insights or speculations about your observations? Have questions arisen based on your observations? Do you have any thoughts on why the collection site provided few or many specimens? Should you consider changing any aspect of your field work protocol?

Freshwater sponges are found in shallow streams and rivers by wading, swimming, and snorkeling. In deeper rivers, lakes, and reservoirs scuba diving or dredging may be required. Scuba diving is the preferred method as dredging can damage benthic environments. As you search for sponges, focus your attention on hard substrates such as rocks, boulders, logs, limbs of fallen trees, bridge abutments, riffraff, shell remains of mollusks, aquatic macrophytes, and discarded human-made objects. View substrates carefully and check the undersides of objects you can lift. Once a sponge has been found it should be checked for the presence of gemmules using a 10x magnifier. Use a scalpel or knife to remove a section of the sponge from its substrate and place the sample into a vial or container containing enough preservative, such as 70% ethanol, to cover the specimen completely. Label each sample with a specific identifier so that it cannot be confused with other samples. Record the identifier on the container and in your field notebook.

Laboratory Techniques

Processed samples should result in clean spicules with little or no debris. To obtain clean spicules for SEM stubs and LM slides, place a portion of sponge-containing gemmules in a test tube and cover with nitric acid or bleach. Under a chemical fume hood, heat the test tube in a hot water bath until the sponge has dissolved leaving only spicules. If a fume hood is not available complete the digestion process in a ventilated room or outdoors and use a fan to blow the acid/bleach fumes away from the technician. Centrifuge the test tube to create a pellet of spicules. Decant the acid and replace it with distilled water, then centrifuge again. Repeat centrifuging, decanting, and replacing the distilled water two more times. Centrifuge a fifth and final time using ethanol.

To prepare for light microscopy (LM) viewing, place a drop or two of the spicule alcohol solution on a glass slide and allow the slide to dry, after which view it under the microscope. To prepare for scanning electron microscopy (SEM) viewing, place a black adhesive carbon tab on the top of a SEM stub. Next, place a circular glass coverslip on top of the carbon tab. Having a glass coverslip under the spicules produces the best results for getting a black background in SEM photographs. Place a drop or two of the spicule alcohol solution on the top of the glass disk. Allow the solution to dry. After drying, sputter coat the stub with gold-palladium. After sputter-coating the stub is ready to be viewed. Carefully view prepared slides and stubs for all classes of spicules. Using a dichotomous key identify the sponge to species or lowest taxon possible.

CHAPTER 5

Taxonomy, Classification, Identification

"What's in a name?"
—WILLIAM SHAKESPEARE—

Taxonomy is the science of grouping biological organisms based on similarities and differences in morphology and genetic relationships. Historically morphology was the only way to classify organisms. Recent advances in biochemical techniques that evaluate genetic relationships have proven more effective and have shown that in some cases groupings of organisms based on morphology got it wrong. There are seven primary levels of taxa (groups): kingdom, phylum, class, order, family, genus, and specific epithet. Primary levels can be subdivided to produce taxa such as suborder or superfamily. Taxa ranking changes occasionally occur. For example, the sponge suborder Spongillina was elevated in rank to a new order Spongillida Manconi and Pronzato, 2002. Table 1 provides the classification of freshwater sponges to family.

There are advantages to using scientific names rather than common names. Common names typically change from one region or country to another resulting in confusion concerning the precise identity of an organism. *Ambloplites rupestris* (Rafinesque, 1817) is a fish with a range extending from northern Alabama to Hudson Bay and from Wisconsin to Delaware. Common names for *Ambloplites rupestris* include rock bass, goggle eye, black perch, redeye, and rock sunfish, among others. The scientific name *Ambloplites rupestris* is the same worldwide, thus among biologists there is no confusion as to the organism's

TABLE 1. Classification of Freshwater Sponges from Kingdom to Family

Taxon	Classification
Kingdom	Animalia
Phylum	Porifera Grant, 1836
Class	Demospongiae Sollas, 1885
Order	Spongillida Manconi and Pronzato, 2002
Family	Lubomirskiidae Rezvoi, 1985 Malawispongiidae Manconi and Pronzato, 2002 Metaniidae Volkmer-Ribeiro, 1986 Metschnikowiidae Czerniavsky,1880 Paleospongillidae Manconi and Pronzato and Reitner, 1991* Potamolepidae Brien, 1967 Spongillidae Gray, 1867

*Extinct

identity. Common names can be misleading and not accurately represent the type of organism they denote. Crayfish are common in the streams of Tennessee, but they are not fish.

Biologists use Latin when naming organisms. The primary reason Latin is used is because during Carolus Linnaeus's life (1707–1778) Latin was the common language used by scientists. Also, Latin is a dead language which does not change and using Latin allows scientists speaking different languages to communicate effectively when discussing organisms.

In the listing of the seven primary taxa, "species" was not included. Why? Because the species name of an organism is written as a binomial, meaning it consists of two terms, genus and specific epithet. This system, of assigning a two-part name to an organism, developed by Carolus Linnaeus, is known as binominal nomenclature.

Consider the freshwater sponge named *Trochospongilla horrida*. *Trochospongilla* is the genus name and *horrida* is the specific epithet. The specific epithet is just that—specific. No other sponge placed in the genus *Trochospongilla* can have the specific epithet *horrida*. However, "horrida" can be used again when naming an organism placed in a different genus. For example, a fish recently discovered in Australian waters was named *Saccogaster horrida* Nielsen,

Schwarzhans and Cohen, 2012. Scientific names are often descriptive of a characteristic of the organism. Consider the freshwater sponge *Cherokeesia armata* Copeland, Pronzato, and Manconi, 2015. The specific epithet *"armata"* is a reference to the gemmular theca being strongly armored by two categories of gemmuloscleres. Also, the epithet may refer to a place, for example *Trochospongilla pennsylvanica* (Potts, 1882) is a reference to the state of Pennsylvania, or the name of an individual as *Trochospongilla leidyi* (Bowerbank, 1863) was named in honor of Joseph Leidy, a famous American paleontologist. Notice that the first letter of the genus name is capitalized and that the entire name is italicized. The binominal name when used for the first time in a document is immediately followed by the name of the person or persons who described and named the organism and the year the name was published. If the name of the person is within parentheses it means the organism has been assigned to a different genus since it was first named. For example, the sponge named *Trochospongilla leidyi* (Bowerbank, 1863) was originally named *Spongilla leidyi* Bowerbank, 1863. A checklist of families, genera, and species of the freshwater sponges of the Nearctic Biogeographical Realm is provided in Table 2.

Sponge identification involves observing micro and macro morphological diagnostic traits. However, biochemical techniques testing for genetic similarities and differences are being used with increasing frequency.

Morphological traits used in classifying freshwater sponges are growth form, consistency, color, surface traits, topographic distribution of inhalant and exhalent apertures, architecture of ectosomal and choanosomal skeleton, topographical distribution and traits of skeletal megascleres, microscleres, gemmuloscleres and gemmules, and gemmular architecture (gemmular cage, foramen, gemmular theca, architecture of pneumatic layer, and special arrangement of spicules).

Spicule morphologies are very important for sponge identification. Of the three-spicule types, gemmuloscleres are the most important taxonomic criterion. An excellent source of illustrations and terminology for describing many sponge diagnostic traits is *Thesaurus of Sponge Morphology* by Boury-Esnault and Rützler published in 1997.

Biologists determine the identity of an organism not known to them by using a dichotomous key. Dichotomous keys are organized as a series of numbered steps. At each step two options are presented of which only one is a true description of some characteristic of the unknown organism. The true statement will

TABLE 2. Checklist of Freshwater Sponge Families, Genera, and Species of the Nearctic Biogeographical Realm

Family	Genus	Species
Metaniidae	Corvomeyenia	*Corvomeyenia carolinensis* Harrison, 1971 *Corvomeyenia everetti* (Mills, 1884)
Potamolepidae	Cherokeesia	*Cherokeesia armata* Copeland, Pronzato and Manconi, 2015
Spongillidae	Anheteromeyenia	*Anheteromeyenia argyrosperma* (Potts, 1880)
	Corvospongilla	*Corvospongilla becki* Poirrier, 1978 *Corvospongilla novaeterrae* (Potts, 1886)
	Dosilia	*Dosilia palmeri* (Potts, 1885) *Dosilia radiospiculata* (Mills, 1888)
	Duosclera	*Duosclera mackayi* (Carter, 1885)
	Ephydatia	*Ephydatia fluviatilis* (L., 1759) *Ephydatia millsi* (Potts, 1887) *Ephydatia muelleri* (Lieberkühn, 1856) *Ephydatia subtilis* Weltner, 1895
	Eunapius	*Eunapius fragilis* (Leidy, 1851)
	Heteromeyenia	*Heteromeyenia baileyi* (Bowerbank, 1863) *Heteromeyenia latitenta* (Potts, 1881) *Heteromeyenia longistylis* Mills, 1884 *Heteromeyenia riojai* Carballo, Gomez, Cruz-Barraza, Yáñez, 2021 *Heteromeyenia tentasperma* (Potts, 1880) *Heteromeyenia tubisperma* (Potts, 1881)
	Heterorotula	*Heterorotula lucasi* Manconi and Copeland, 2022
	Pottsiela	*Pottsiela aspinosa* Potts, 1880
	Racekiela	*Racekiela biceps* (Lindenschmidt, 1950) *Racekiela cresciscrystae* Gómez, Carballo, Cruz-Barraza, and Camacho-Cancinoa, 2019 *Racekiela montemflumina* Carballo, Cruz-Barraza, Yáñez, and Gómez, 2017 *Racekiela pictouensis* (Potts, 1885) *Racekiela ryderi* (Potts, 1882)
	Radiospongilla	*Radiospongilla cerebellata* (Bowerbank, 1863) *Radiospongilla crateriformis* (Potts, 1882)
	Spongilla	*Spongilla alba* Carter, 1849 *Spongilla cenota* Penney and Racek, 1968 *Spongilla lacustris* (L., 1759) *Spongilla wagneri* Potts, 1889
	Stratospongilla	*Stratospongilla penneyi* (Harrison, 1979)
	Trochospongilla	*Trochospongilla horrida* (Weltner, 1893) *Trochospongilla leidyi* (Bowerbank, 1863) *Trochospongilla pennsylvanica* (Potts, 1882)

direct you to the next step in the key. As you move from one step to the next the number of possible identities for a specimen is reduced. Keep following the true statements until you arrive at the name of the organism. A dichotomous key for the sponges found in Tennessee is provided at the end of chapter 6 and a list of freshwater sponges for each state can be found in appendix B.

CHAPTER 6

Sponges of Tennessee

*"The mind is not a vessel to be filled,
but a fire to be kindled."*

—PLUTARCH—

When describing life-forms, biologists use terminology many people find difficult to understand. Do not be intimidated by the terminology. The SEM photographic plates of each species provide a visual representation of the descriptive terms. By studying the plates as you read the descriptions of spicules and gemmules, you will quickly gain an understanding and begin using these terms as you discuss sponges with your colleagues.

Three families of freshwater sponges occur in the United States: Metaniidae, Potamolepidae, and Spongillidae. Two families, ten genera, and fourteen species are currently known to occur in Tennessee (Table 3).

Family and species descriptions along with range maps of the freshwater sponges of Tennessee are presented in the following pages. Species descriptions are from Penney and Racek (1968), Reiswig, et al. (2010), Manconi and Pronzato (2016) and from relevant journal publications.

In the following descriptions, all measurements are from Tennessee specimens. Ranges for gemmule diameters, spicular lengths, and rotule diameters are provided, followed by mean and standard deviation in parentheses. Precise collection sites cannot be determined on range maps of this scale. However, the black circles on these maps do indicate the general area from which a sponge

Stephanie Williams, Tennessee Department of Environment and Conservation; Bill Reeves, Tennessee Wildlife Resources Agency; and David Withers, Tennessee Department of Environment and Conservation, at Richland Creek, Davidson County, Tennessee.

species was collected. The names of rivers and drainages from which sponges were collected can be determined.

Family and Species Descriptions

Potamolepidae

Family Description (from Copeland, et al., 2015): Spongillida with **growth form** encrusting or massive to arborescent with irregular lobes, ridges, or branches. **Consistency** is rigid to stony hard, fragile at the sponge base which typically remains attached to the substrate when collected. **Surface** smooth to irregular with chimneys, to conulose (cone-like) with apices of ascending primary fibers supporting conules (cone-shaped). **Oscules** with exhalent star shaped

TABLE 3. Families, Genera, and Species of the
Freshwater Sponges Occurring in Tennessee

Family	Genus	Species
Potamolepidae	*Cherokeesia*	*Cherokeesia armata*
Spongillidae	*Corvospongilla*	*Corvospongilla becki*
	Ephydatia	*Ephydatia fluviatilis*
		Ephydatia muelleri
	Eunapius	*Eunapius fragilis*
	Heteromeyenia	*Heteromeyenia latitenta*
		Heteromeyenia tubisperma
	Heterorotula	*Heterorotula lucasi*
	Racekiela	*Racekiela ryderi*
	Radiospongilla	*Radiospongilla cerebellata*
		Radiospongilla crateriformis
	Spongilla	*Spongilla lacustris*
	Trochospongilla	*Trochospongilla horrida*
		Trochospongilla leidyi

canals in some cases. No special **ectosomal skeleton** except for spicular tufts supporting conules and tangential microscleres in the dermal membrane (only in some genera). **Spongin** is sparse except for the basal spongin plate and the gemmular theca (case or capsule portion of gemmule that surrounds archeocytes). **Choanosomal skeleton** irregular alveolate-reticulate (pitted and lattice-like) with mono- to paucispicular tracts sometime ascending toward the surface, skeletal network notably dense at the surface and loose and irregular at the sponge base. **Megascleres** strongyles (spicules rounded at both ends) to oxeas (spicules pointed at both ends) from smooth to ornamented by variably dense granules/spines/tubercles. **Microscleres** when present, slender oxeas. **Gemmules** when present are single to grouped, sometimes free at the sponge base or more frequently sessile and adhering to the substratum by the sponge basal plate. **Gemmular theca** mono-, bi- to tri-layer, usually of compact spongin with more or less tangentially embedded gemmuloscleres. The **Pneumatic layer** (air chambers or mesh layer) is absent to scarcely developed and, if present, of fibrous, not chambered spongin. **Gemmuloscleres** strongyle-like, short to elongate, ovular to variably bent (c-shaped to button-like) variably ornamented to entirely smooth. A second class of gemmuloscleres as large, stout, spiny to smooth oxeas. **Parenchymella** larvae entirely ciliated, with spicules.

CHEROKEESIA ARMATA COPELAND, PRONZATO, AND MANCONI, 2015

Species Description: From Copeland et al., (2015). **Color** in vivo (living): light gray somewhat transparent, becoming white in alcohol. **Growth form** (Fig. 34) is encrusting as minute cushions, 1–4 mm in thickness, up to at least 10 cm in diameter, **Consistency** is hard but fragile in vivo. **Surface** is alveolate (pitted) with slight hispidation (spiny) by more or less erected ectosomal spicules supporting small scattered conules. **Oscules** conspicuous in vivo, numerous and scattered. **Ostia** are scattered. **Ectosomal skeleton** irregularly alveolate as a mesh network of mono- to paucispicular tracts (oxeas) supporting the dermal membrane. **Choanosomal skeleton** irregularly alveolate, with monospicular polygonal meshes (100–200 μm in diameter) a few, scarcely developed,

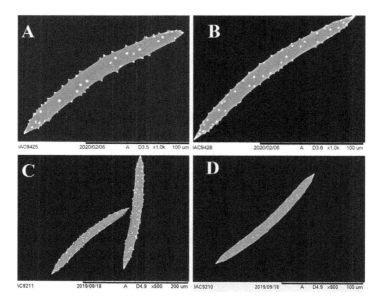

PLATE 1. *Cherokeesia armata*: Megascleres. **A-D**, Stout spiny to nearly smooth acanthoxeas.

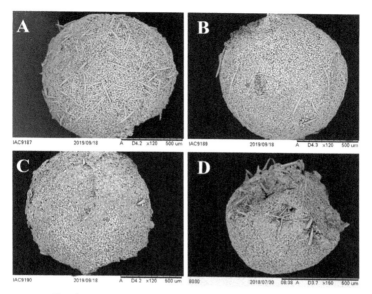

PLATE 2. *Cherokeesia armata*: Gemmules. **A**, **B**, **C**, and **D**, Showing gemmuloscleres and very thin outer layer of spongin; **D**, notice foramen at the 3 o'clock position.

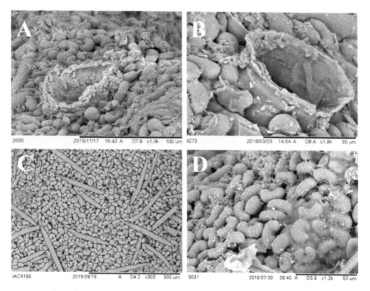

PLATE 3. *Cherokeesia armata*: Gemmular surface. **A**, closed foramen having a short, simple collar; **B**, open foramen; **C**, surface of gemmules showing large spiny acanthoxea gemmuloscleres and smooth strongyle-like gemmuloscleres; **D**, close-up of thin outer layer of spongin and smooth strongyle-like gemmuloscleres.

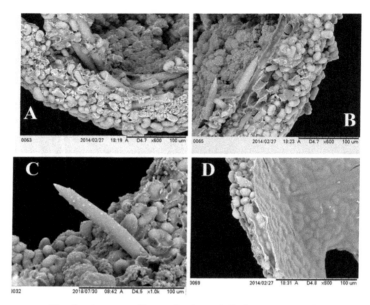

PLATE 4. *Cherokeesia armata*: Gemmular theca. **A-B**, Cross sections showing multilayer arrangement of gemmuloscleres and staminal cells; **C**, large spiny acanthoxea gemmulosclere surrounded by stronglye-like gemmuloscleres; **D**, thin wall section of gemmular theca showing inner membrane with impressions of gemmuloscleres and foraminal aperture.

ascending paucispicular tracts. **Spongin** notably scanty, except for the gemmular theca and the basal spongin plate. **Basal spongin plate** is well-developed and adhering to the flat base of the gemmules. **Megascleres** stout spiny straight to slightly bent oxeas 161.8–228 (199.5 ± 15.8) µm in length, with gradually pointed to abruptly pointed tips; spines straight to recurring toward the tip. **Microscleres** are absent. **Gemmules** 591–957 (749 ± 84) µm in diameter, sessile at the sponge base, hemispherical in shape, strictly adhering to the substrate by the basal spongin plate. Gemmules occurring single or in clusters of 3–4 and strongly armed with gemmuloscleres of two types. **Foramen** single to multiforamina (up to 3) having a simple collar 20–40 µm in diameter which may be closed by a spongin lamina. **Gemmular theca** of compact spongin variably thick (25–150 µm), trilayered, with up to 10 layers of gemmuloscleres of

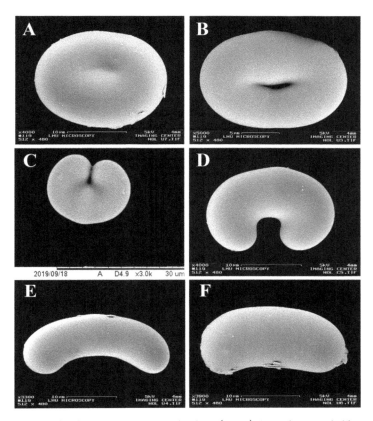

PLATE 5. *Cherokeesia armata*: Gemmuloscleres (type I). Smooth strongyle-like, A-F, sequence of development from circular to C-shape to ovoid strongyles.

two categories intermingled and tangentially embedded. **Outer layer** of thin compact spongin, like a film on gemmuloscleres surfaces. **Middle layer** of a thin non-alveolate layer of compact spongin (pneumatic layer absent). **Inner layer** of compact spongin. **Gemmuloscleres** of two types (strongyle-like spicules and spiny to smooth oxeas). **Type 1**. Dominant in several layers (5–7) more or less in a mosaic-manner, joined by a conspicuous amount of compact spongin. Strongyle-like gemmuloscleres 17–29.3 (22.4 ± 2.5) μm in length, stout, entirely smooth, slightly bent, to extremely bent C shape to bean shape to button-like.

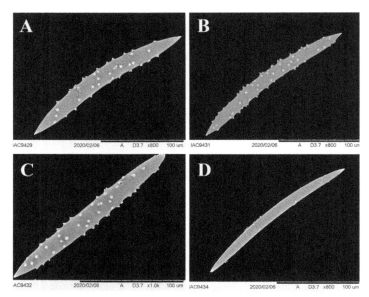

PLATE 6. *Cherokeesia armata*: Gemmuloscleres (type II). **A-D,** Spiny to nearly smooth acanthoxeas, similar in appearance to megascleres.

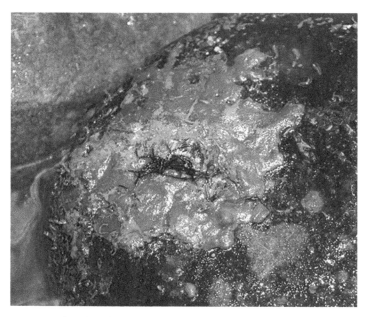

FIG. 34. *Cherokeesia armata*

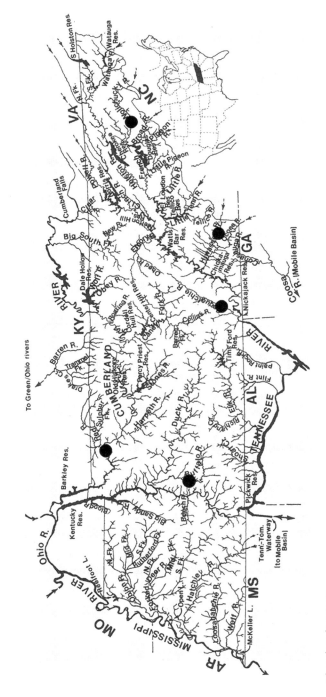

Cherokeesia armata.

Rare, elongate, short typical strongyles, slightly bent to straight 50–100 μm in length. **Type II.** Less abundant, large, stout spiny to smooth oxeas tangentially embedded intermingled with gemmuloscleres type I in the entire thickness of theca. Oxeas 173.3–258 (219.7 ± 16.7) μm in length, large, stout, spiny to smooth with acute straight to recurved spines toward the tips.

Cherokeesia armata represents the only known living member of the family Potamolepidae in the Northern Hemisphere. Until the discovery of *C. armata* in Tennessee the living members of Potamolepidae were known to have a Southern Hemisphere distribution in the tropical regions of Neotropical, Afrotropical, Australian, Oriental and Pacific Island biogeographical regions. Two fossil members of Potamolepidae have been documented in the Northern Hemisphere. *Oncosclera kaniensis* Matsuoka and Masuda, 2000, discovered from the Miocene in Japan and *Potamophloios canadensis* Pisera, Siver, and Wolf, 2013 from the middle Eocene of northern Canada.

Cherokeesia armata has a Nearctic geographical distribution. Currently known only from Tennessee.

Spongillidae

Family Description (Mancini and Pronzato, 2002): **Shape** ranging from globular, to massive, encrusting, lobate, arborescent. **Color** in vivo white to brown, green with alga symbiont, from pale to dark, **Surface** smooth, hispid, conulose. **Consistency** from very soft to rigid. **Ectosomal skeleton** from spicular brushes at apices of primary fibers to dense tangential spicular network. **Choanosomal skeleton** reticulate with regular to irregular meshes. **Megascleres** are smooth to spiny or granular oxeas or strongyles. **Microscleres** if present oxeas, strongyles, asterlike (star-shaped) pseudobirotules. **Gemmules** present except in perennial specimens. **Gemmular cage** composed of megascleres present in several species. **Foramina** are present except in the genus *Nudospongilla*. **Gemmular theca** mono-, bi-, or trilayered. Gemmules are usually armed with gemmuloscleres but in some cases naked. **Gemmuloscleres** partially or totally embedded in gemmular theca, arranged radial or tangential to the surface. **Pneumatic layer** present or absent. Gemmuloscleres of various forms and in many cases species-specific, from spiny to smooth, from oxeas to strongyles, birotules (having a rotule at each end of the spicule), pseudorotules (having false rotules) club-like, botryoidal (like a clump of grapes). **Parenchymella** larvae.

Fourteen species of Spongillidae have been documented from Tennessee waters.

CORVOSPONGILLA BECKI POIRRIER, 1978

Species Description: From Poirrier (1978); Manconi and Pronzato (2016). **Color** in vivo: dark gray to black. **Consistency** firm. **Surface** smooth, even, having subdermal canal network. **Oscula** conspicuous in vivo, with subdermal radial canal. **Ostia** scattered. **Ectosomal skeleton** with abundant tangential microscleres and tips of ascending fibers to support the dermal membrane. **Choanosomal skeleton** reticulate network having mono- to paucispicular tracts and scattered microscleres. **Spongin** scanty, except for gemmular theca and basal spongin plate. **Basal spongin plate** is well-developed, armed by strongyles with forms intermediate between megascleres and gemmulosclentes. **Megascleres** stout pointed amphioxea (slightly curved needlelike spicule) to amphistrongyla (slightly curved spicule with rounded ends) 125–206.7 (173.2 ± 20.2) μm in length which are microspined with larger spines on the ends. Megascleres are mostly curved, a few straight, and widest in the middle. Some megascleres of both types with a single elongated spine at the tips. Spicules which are morphologically intermediate in size between megasclere and

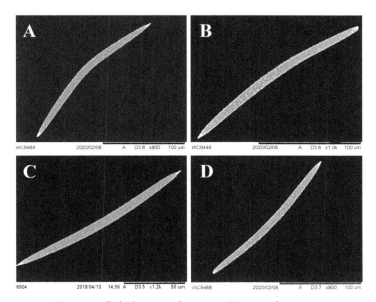

PLATE 7. *Corvospongilla becki*: Megascleres. **A-D**, Spiny amphioxeas.

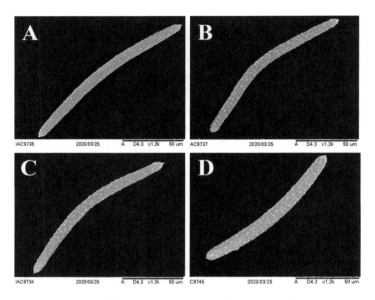

PLATE 8. *Corvospongilla becki*: Megascleres. **A-D**, Spiny strongyles.

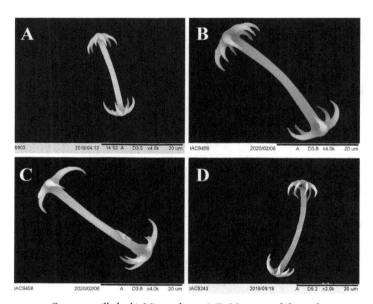

PLATE 9. *Corvospongilla becki*: Microscleres. **A-D**, Micropseudobirotules

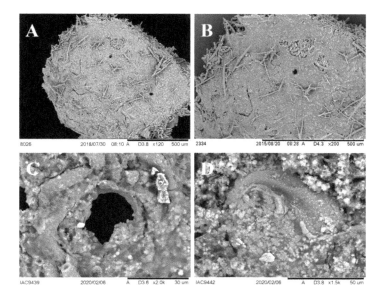

PLATE 10. *Corvospongilla becki*: Gemmules. **A**, Two attached gemmules each having an open foraminal aperture; **B**, surface cover by spongin and spicules, with open foraminal aperture; **C**, close-up view of open foramen; **D**, close-up of closed foramen.

gemmulosclere amphistrongyla are found among the gemmules in the basal layer. **Microscleres** are micropseudobirotules 12.9–54.5 (33.2 ± 7.1) μm in length with smooth, slightly curved to straight shafts, having 4–6 recurved hooks 5.1–15 (10.98 ± 2.1) μm in diameter at ends. **Gemmules** 635–1007 (880 ± 140) μm in diameter, are subspherical, occurring in several layers and firmly attaching to the substrate. **Gemmular theca** with **outer layer** spongin bearing longer tangential gemmuloscleres and irregularly embedded shorter gemmuloscleres in a mosaic-like pattern. **Foramen** is a short tube, usually single but may find as many as 2–4. **Pneumatic layer** is absent. **Inner layer** of compact spongin with sublayers. **Gemmuloscleres** are amphistrongyles of two size groups. Small gemmuloscleres 26.4–64.2 (40.9 ± 7.2) μm in length, slightly to strongly curved having microspines except in the inner curved region, larger gemmuloscleres 70.5–139.3 (97.2 ± 17.3) μm in length, are straight, microspined and tuberculated. Gemmuloscleres are arranged irregularly in the gemmular coat. Larger gemmuloscleres are tangential arranged intermixed with smaller

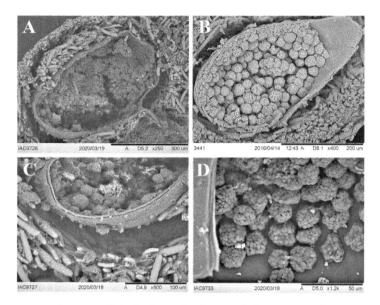

PLATE 11. *Corvospongilla becki*: Gemmular theca. **A-C**, Cross sections showing the absence of a pneumatic layer and layering of inner membrane, with clusters of staminal cells; **D**, close-up of staminal cells.

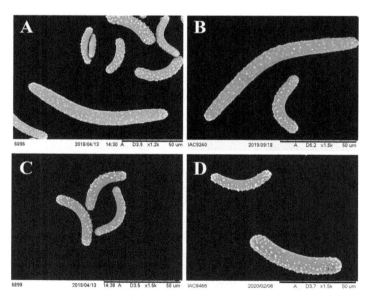

PLATE 12. *Corvospongilla becki*: Gemmuloscleres. **A-D**, Spiny amphistrongyles of two sizes (notice the scarcity of spines in the curved sections of the gemmuloscleres.

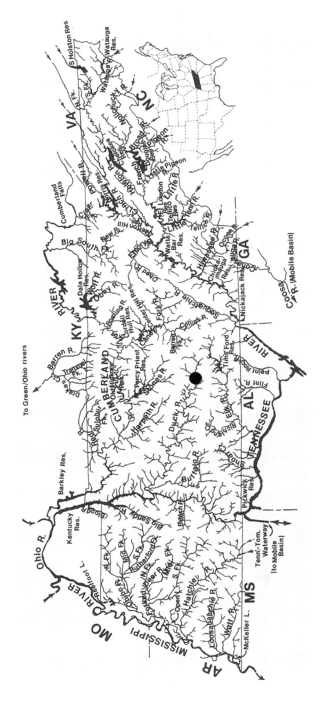

Corvospongilla becki.

gemmuloscleres which are arranged at irregular angles. This arrangement of gemmuloscleres results in a compact assemblage of closely packed spicules.

Corvospongilla becki is endemic to the Nearctic Region and has a spotty distribution in the United States. It is reported from the type locality of Duck Lake in the Atchafalaya Basin, St. Martin Parish, Louisiana, the Cahaba River and Shades Creek of Alabama. In Tennessee only two specimens of this sponge have been collected, both from the Duck River at Henry Horton State Park. *Corvospongilla becki* may be the rarest sponge in Tennessee.

EPHYDATIA FLUVIATILIS (LINNAEUS, 1759)

Species Description: **Color** in vivo: drab yellow to brown, green due to green alga symbiont. **Growth form** (Fig. 35) is encrusting, bulbous, may be massive. **Consistency** firm but fragile in vivo, extremely brittle when dry. **Ectosomal skeleton** no special architecture. **Choanosomal skeleton** is anisotropic (having unequal values when measured in different directions) with paucispicular

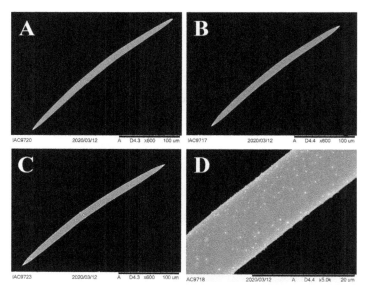

PLATE 13. *Ephydatia fluviatilis*: Megascleres. **A-C**, Slightly bent to straight, smooth oxeas; **D**, surface of megasclere with microspines.

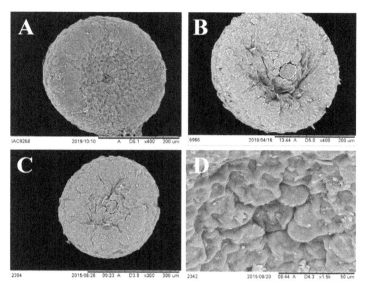

PLATE 14. *Ephydatia fluviatilis*: Gemmules. **A-C**, Covered in spongin with impressions of gemmuloscleres; **D**, close-up of closed gemmular foramen (in center) among gemmulosclere impressions.

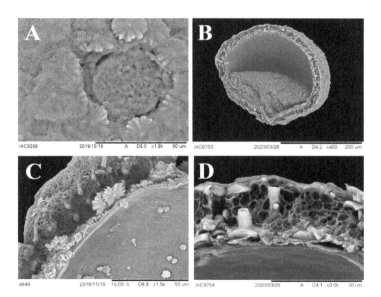

PLATE 15. *Ephydatia fluviatilis*: Foramen and cross section of gemmular theca. **A**, Close-up of closed foramen; **B**, cross section of gemmular theca showing staminal cells, radial arrangement of gemmuloscleres; **C-D**, close-up of gemmular thecae showing broken gemmuloscleres, chambered pneumatic membrane, and inner membrane of compact multilayered spongin.

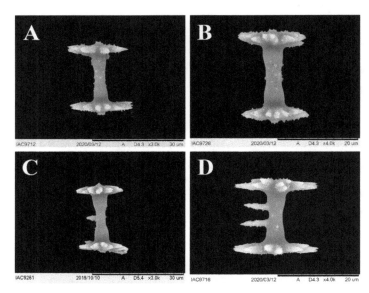

PLATE 16. *Ephydatia fluviatilis*: Gemmuloscleres. **A-D**, Birotules having flat incised, microspined, rotules of equal diameter; nonspiny (**A** and **B**) to spiny (**C** and **D**) shafts.

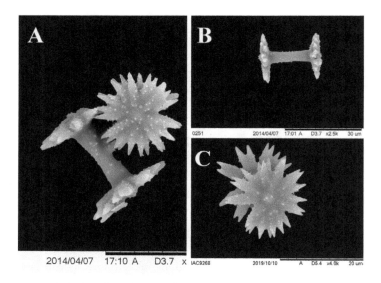

PLATE 17. *Ephydatia fluviatilis*: Birotules. **A-C**, margins incised and microspined.

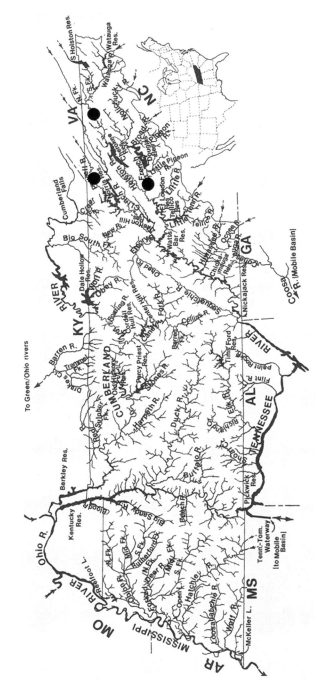

Ephydatia fluviatilis.

fibers and tracts (fiber or tract with 2–5 megascleres adjacent to one another). **Spongin** scanty except for gemmular theca. **Megascleres** are smooth to microspined oxeas 234–332 (296.4 ± 22.4) µm in length. **Microscleres** are absent. **Gemmules** 322–425 (386 ± 35) µm in diameter, subspherical, scattered to grouped in carpets at sponge basal portion. **Foramen** simple, slightly elevated with collar. **Gemmular theca** trilayered. The **Outer layer** well developed spongin frequently covering gemmuloscleres. **Pneumatic layer** having irregular rounded chambers with gemmuloscleres arranged radially in one layer. **Inner layer** compact spongin arranged in sublayers, in contact with gemmuloscleres. **Gemmuloscleres** are birotules of one length group 18.8–26.1 (22.5 ± 1.8) µm in length having smooth shafts with 0–4 large spines, with flat rotules of equal diameters 14–23.2 (18.7 ± 1.8) µm in diameter, having more than 20 shallowly incised teeth.

Ephydatia fluviatilis is widely distributed within northern and southern hemispheres. The geographical distribution of *E. fluviatilis* includes the Ne-

FIG. 35. *Ephydatia fluviatilis*

Ephydatia fluviatilis, Indian Creek, Claiborne County, Tennessee.

arctic, Palearctic, Oriental, Australian, and Afrotropical Regions. It has been documented from Europe, Russia, China, Japan, India, Australia, Africa, and North America. *Ephydatia fluviatilis* is one of a few freshwater sponges which produce gemmules capable of withstanding long periods of drought. Reported to prefer cold to warm temperate regions (Penney and Racek, 1968).

EPHYDATIA MUELLERI (LIEBERKÜHN, 1855)

Species Description: **Color** in vivo: drab yellow to brown, green due to green alga symbiont. **Growth form** (Fig. 36) encrusting or thin cushion. **Consistency** in vivo firm and moderately hard. **Oscula** scattered. **Surface** slightly hispid. **Ectosomal skeleton** with no special architecture. **Choanosomal skeleton** anisotropic with primary ascending fibers joined by vague secondaries. **Spongin** scanty except for gemmular theca. **Megascleres** 183–301 (246.2 ± 30) µm in length, slightly curved to straight stout oxeas, except for tips covered with small spines, rarely entirely smooth. **Microscleres** are absent. **Gemmules**

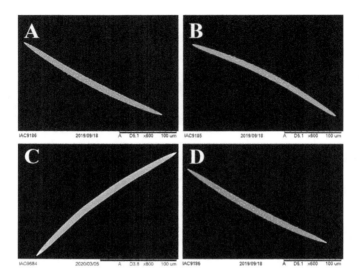

PLATE 18. *Ephydatia muelleri*: Megascleres. **A-D**, Straight to slightly curved acanthoxeas; **A**, **B**, and **D**, spiny except at tips; **C**, smooth.

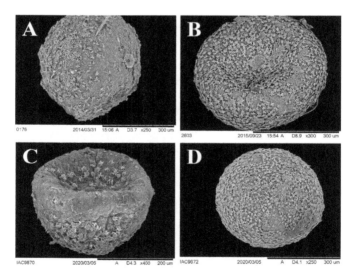

PLATE 19. *Ephydatia muelleri*: Gemmules. **A-D**, Outer membrane of spongin with protruding gemmuloscleres.

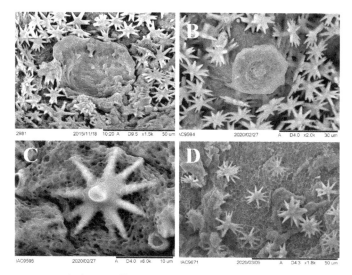

PLATE 20. *Ephydatia muelleri*: Gemmular surface. **A-B**, Close-up of closed foramina; **C**, deeply incised toothed, ray-like rotule covered by spongin; **D**, gemmular surface of spongin and protruding gemmuloscleres.

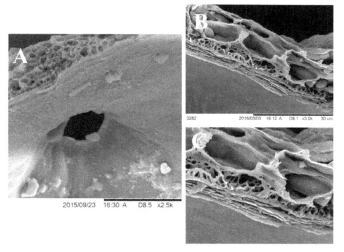

PLATE 21. *Ephydatia muelleri*: Gemmular theca. **A**, (Looking outward from within the gemmule) inner membrane with open foramen and chambered pneumatic layer; **B-C**, close-up of gemmular theca showing multilayered inner membrane with thin pneumatic layers.

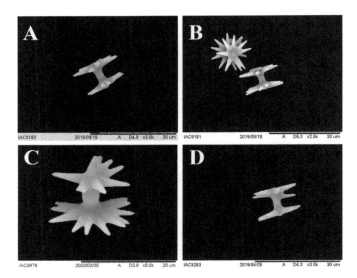

PLATE 22. *Ephydatia muelleri*: Gemmuloscleres. **A-D**, Birotules having short, stout shafts bearing flat, deeply incised toothed rotules; **B-C**, rotules nonuniform in shape.

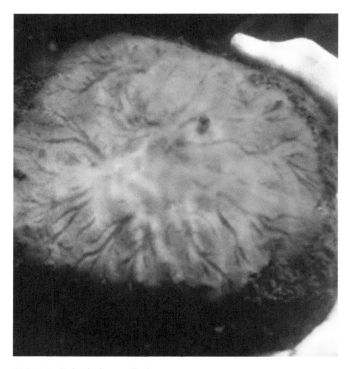

FIG. 36. *Ephydatia muelleri*

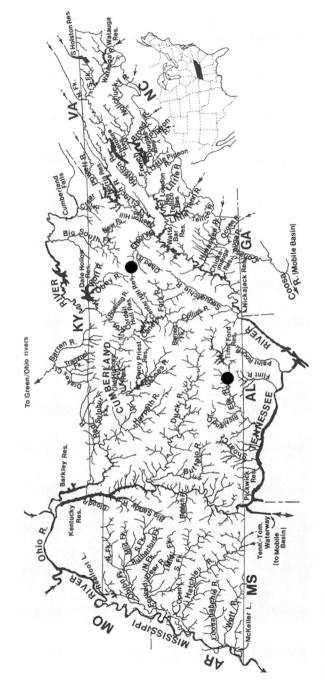

Ephydatia muelleri.

349–460 (400 ± 39) μm in diameter, subspherical, abundant, scattered in body or may aggregate in the basal portion without forming a pavement layer. **Foramen** slightly elevated with flattened collar. **Gemmular theca** trilayered. **Outer layer** of compact spongin scarcely developed. **Pneumatic layer** of chambered spongin having small, rounded (more or less) chambers and 1–4 layers of gemmuloscleres radially embedded. **Inner layer** of sublayered compact spongin. **Gemmuloscleres** birotules of one size class 8.3–15.3 (11.2 ± 1.2) μm in length, having a short shaft, rotules equal in diameter 10.8–23 (16.9 ± 2.1) flattened and deeply incised to form usually 12 or fewer long rays.

Ephydatia muelleri has a wide distribution. It's reported from the Nearctic, Palearctic, Afrotropical, and Neotropical Regions.

EUNAPIUS FRAGILIS (LEIDY, 1851)

Species Description: **Color** in vivo: light gray to whitish, green due to green alga symbiont. **Growth form** (Fig. 37) encrusting, variably thick. **Consistency** soft and extremely fragile. **Surface** slightly hispid. **Oscules** not conspicuous. **Ectosomal skeleton** without special architecture. **Choanosomal skeleton**

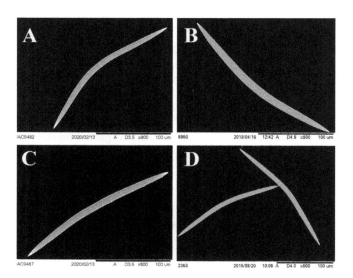

PLATE 23. *Eunapius fragilis*: Megascleres. **A-D**, Smooth, straight to curved oxeas.

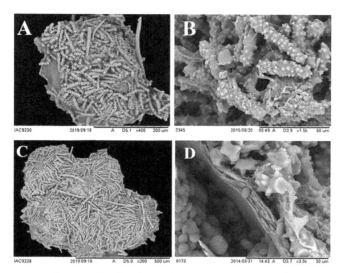

PLATE 24. *Eunapius fragilis*: Gemmules. **A**, View of outer membrane, gemmules, and foramen; **B**, gemmule surface of gemmuloscleres and open foramen; **C**, cluster of five gemmules; **D**, gemmular theca cross section showing outer membrane covering of gemmuloscleres and slight amount of spongin, inner membrane, and staminal cells.

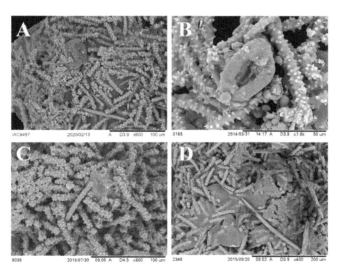

PLATE 25. *Eunapius fragilis*: Gemmular foramina and surface. **A-C**, Gemmular foramina surround by gemmuloscleres; **D**, surface of gemmule showing outer membrane of spongin and gemmuloscleres.

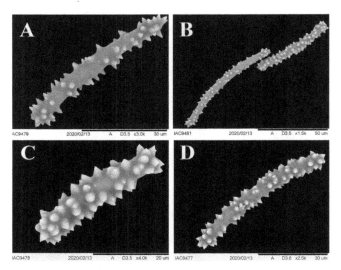

PLATE 26. *Eunapius fragilis*: Gemmuloscleres. **A-D**, Straight to slightly curved spiny strongyles.

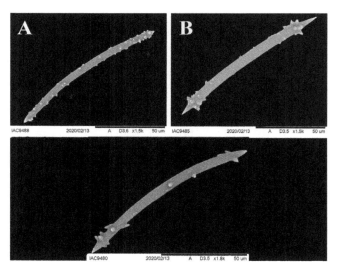

PLATE 27. *Eunapius fragilis*: Gemmuloscleres. **A-D**, Straight to slightly curved sparsely spined oxeas.

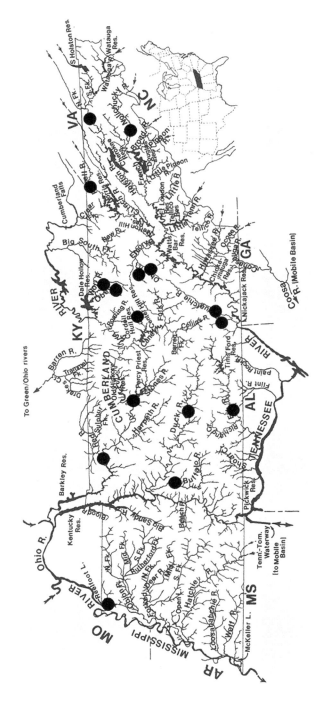

Eunapius fragilis.

FIG. 37. Eunapius fragilis

irregular reticulate network. **Spongin** scanty except for gemmular theca and basal spongin plate. **Megascleres** are smooth, slightly curved to straight fusiform oxeas 154–251 (199 ± 19.8) μm in length. **Microscleres** are absent. **Gemmules** 337–481 (376 ± 54) μm in diameter, when mature are brown in color, either forming a pavement layer cemented to the substrate or in clusters of 2–4 gemmules. **Foramen** tubular. **Gemmular theca** trilayered. **Outer layer** thin layer of compact spongin. **Pneumatic layer** rounded chambers, with one to four layers of gemmuloscleres tangentially embedded. **Inner layer** compact laminar spongin with sublayers. **Gemmuloscleres** slightly curved to straight strongyles 35–113 (68.2 ± 14.8) μm in length, covered with spines, and a few oxeas having sparse spines.

Eunapius fragilis is a cosmopolitan distributed species. It has been reported from all zoogeographic regions except for Antarctica. *Eunapius fragilis* is broadly distributed across Tennessee.

HETEROMEYENIA LATITENTA (POTTS, 1881)

Species Description: **Color** in vivo: yellowish, green due to green alga symbiont. **Growth form** cushion-like, small. **Consistency** in vivo soft, loose texture. **Surface** hispid due to more or less erect spicules. **Oscula** not conspicuous. **Ostia** scattered. **Ectosomal skeleton** without special architecture. **Choanosomal skeleton** irregular network of ascending fibers. **Megascleres** are slender, smooth to sparsely covered with small spines, fusiform oxeas 282–377 (334.5 ± 23.1) μm in length. **Microscleres** completely spined slender oxeas 10.2–14.5 (12.5 ± 1.1) μm in length. **Gemmules** 420–483 (450 ± 23) μm in diameter, subspherical. **Foramen** tubular, long having one or two cirrous projections extending from a flat disk. **Gemmular theca** trilayered, with radially embedded gemmuloscleres. **Outer layer** with protruding gemmuloscleres. **Pneumatic layer** spongin well-developed with a single layer of gemmuloscleres radially embedded. **Inner layer** of compact spongin with sublayers. **Gemmuloscleres** are birotules 44.8–69.2 (55.8 ± 7.4) μm in length, found in one broad overlapping length group, or in some populations of two length sizes. Shafts with long

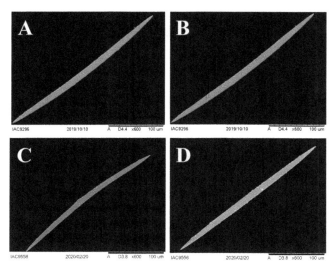

PLATE 28. *Heteromeyenia latitenta*: Megascleres. **A-D**, Oxeas, straight, smooth to sparsely covered by microspines except near tips.

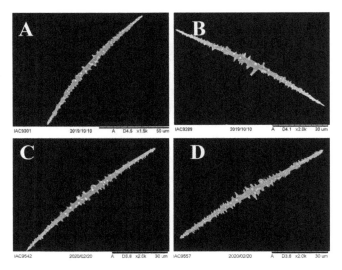

PLATE 29. *Heteromeyenia latitenta*: Microscleres. **A-D**, Oxeas, slender, entirely spined, spines pointed near tips but becoming larger and blunt in middle region of spicule.

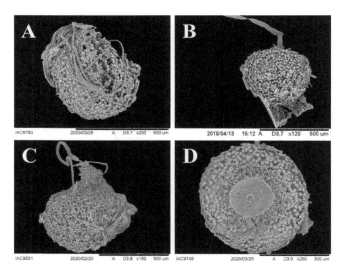

PLATE 30. *Heteromeyenia latitenta*: Gemmules. **A-C**, With single foramen having long ribbon-like cirrus, gemmuloscleres prominent, scanty spongin; **D**, flat disk missing its cirrus.

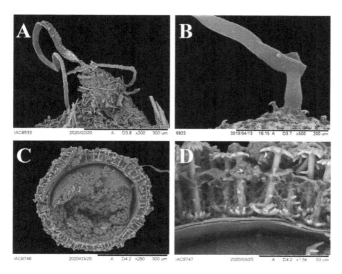

PLATE 31. *Heteromeyenia latitenta*: Foramina and Gemmular Theca. **A-B**, Close-up views of foramina and long cirrus; **C**, cross section of gemmular theca, (from inside out) staminal cells, inner membrane, pneumatic layer, gemmuloscleres, and outer layer of spongin; **D**, close-up of thecal cross section showing single layer of gemmuloscleres arranged radially and embedded in pneumatic layer, and inner membrane.

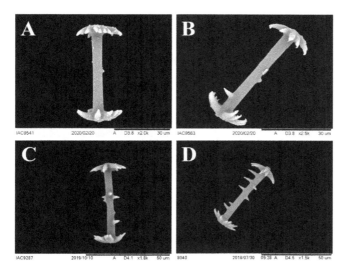

PLATE 32. *Heteromeyenia latitenta*: Gemmuloscleres. **A-D**, Umbonate birotules with shafts having few to numerous sharp-pointed spines.

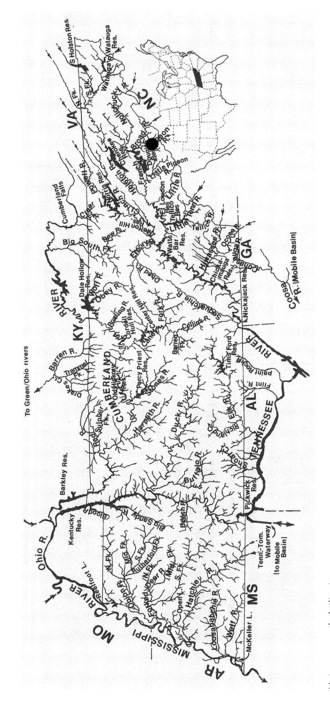

Heteromeyenia latitenta.

stout spines. **Rotules** are of equal size 16.8 –28.1 (21 ± 2) μm in diameter, with numerous recurved hooks.

This species was the second most frequently collected sponge from the Pigeon River of eastern Tennessee. The Pigeon River has a long history of dioxin pollution. Fortunately, by the mid 1990s the Pigeon River was restored to the point where reintroduction of extirpated species occurred, none of which were sponges. The paper mill responsible for pollution ceased production in 2023.

HETEROMEYENIA TUBISPERMA (POTTS, 1881)

Species Description: **Color** in vivo: yellowish, green due to green-algae symbiont. **Consistency** in vivo soft. **Growth form** (Fig. 38) massive, encrusting. **Surface** uneven to slightly papillose. **Oscula** not conspicuous. **Ostia** scattered irregularly. **Ectosomal skeleton** without special architecture. **Choanosomal skeleton** irregular of ascending fibers joined by secondaries and scattered microscleres. **Megascleres** slender, sharp pointed and sparsely microspined oxeas 216–322 (288 ± 19.6) μm in length. **Microscleres** long slender oxeas 66–134 (108.7 ± 14.4) μm in length, entirely spined with spines in center of shaft

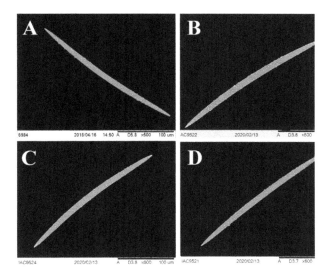

PLATE 33. *Heteromeyenia tubisperma*: Megascleres. **A-D**, Slender, nearly smooth to sparsely microspined except near tip oxeas.

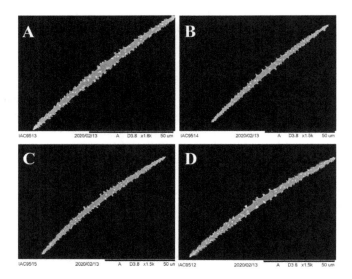

PLATE 34. *Heteromeyenia tubisperma*: Microscleres. **A-D**, Oxeas which are sharp-pointed, slightly curved to nearly straight with large spines some as rosettes.

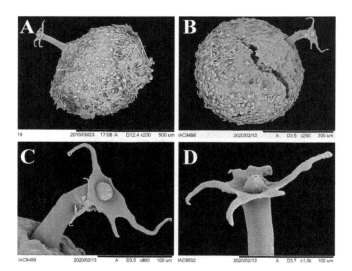

PLATE 35. *Heteromeyenia tubisperma*: Gemmules. **A-B**, Entire gemmules with foramen and outer membrane; **C-D**, close-up of foramen with long terminal cirri.

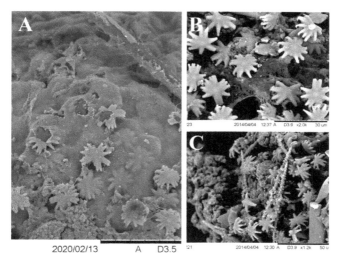

PLATE 36. *Heteromeyenia tubisperma*: Gemmular surface. **A**, Outer surface of spongin with protruding rotules of gemmuloscleres; **B**, close-up of protruding rotules of gemmuloscleres; **C**, microscleres on gemmular surface.

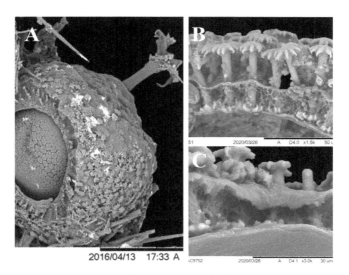

PLATE 37. *Heteromeyenia tubisperma*: Gemmule theca. **A**, Cross section showing thesocytes surrounded by gemmular theca; **B**, tri-layered theca showing pneumatic layer with proximal rotules of gemmuloscleres laying on surface of inner membrane and distal rotules above outer membrane; **C**, close-up of compact spongin inner and outer membranes and pneumatic layer.

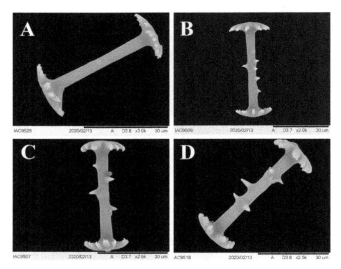

PLATE 38. *Heteromeyenia tubisperma*: Gemmuloscleres. **A**, Smooth shaft, rotules; **B-D**, shafts variously spined, rotules of equal diameters and spined laterally.

FIG. 38. *Heteromeyenia tubisperma*

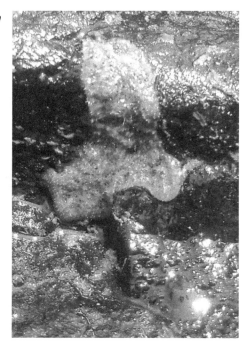

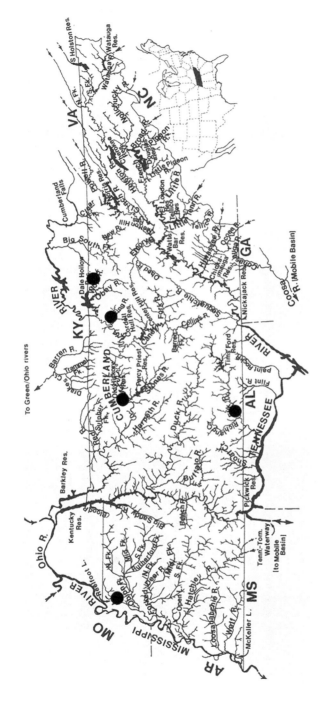

Heteromeyenia tubisperma.

distinctly longer than those toward the tips. **Gemmules** 483–535 (512 ± 16) μm in diameter, subspherical. **Foramen** of gemmule slender and long bearing 4–10 cylindrical cirri (tendril-like structures found on the foramen apertures of sponges of the genus *Heteromeyenia*) projections. **Gemmule theca** trilayered. **Outer layer** of compact spongin with gemmulosclseres protruding. **Pneumatic layer** having small, rounded chambers, gemmulosclseres arranged in a single radial layer. **Inner layer** of compact spongin with sublayers. **Gemmulosclseres** are birotules 45–69 (54.8 ± 6.5) μm in length, of one size group. Shafts of gemmulosclseres with a few spines to entirely smooth; rotules equal in size 14–26 (20.5 ± 2.6) μm in diameter, with numerous deeply cleft recurved teeth.

Current distribution of *H. tubisperma* is the eastern half of North America, from Louisiana into Canada.

HETEROROTULA LUCASI MANCONI AND COPELAND, 2022

Species Description: **Color** in vivo: tawny. **Consistency** of live sponge soft and fragile. Spongin scanty in skeletal network, arranged as irregular polygonal meshes, to abundant in gemmular theca, and basal spongin plate. **Growth form** (Fig. 39) encrusting 1–4 mm in thickness and 16 cm in diameter. **Surface** with irregular ridges, hispid from emerging spicules, and with a network of subdermal canals covered by hyaline dermal membrane. **Ectosomal skeleton** of slender megascleres in paucispicular fibers, with no special architecture

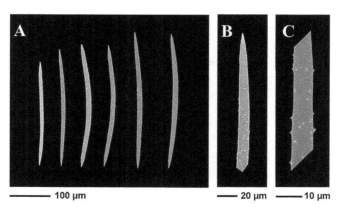

PLATE 39. *Heterorotula lucasi*: Megascleres. **A**, Fusiform acanthoxeas, from nearly spineless to variable-spined except toward the tip; **B**, smooth tip of megasclere; **C**, spiny shaft of megasclere.

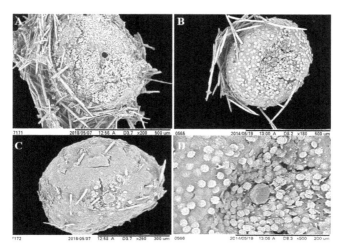

PLATE 40. *Heterorotula lucasi*: Gemmules. **A**, Old gemmule after hatching with open foramen, gemmule surrounded by spicular cage of acanthoxeas; **B**, gemmule with partial gemmular cage and outer layer armed by dense distal rotules of gemmuloscleres; **C**, gemmule with slightly elevated closed foramen, outer membrane of spongin, **D**, fibrous outer gemmular membrane, foramen elevated slightly and closed by a membrane.

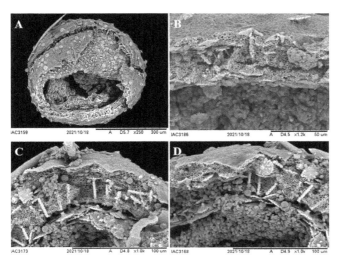

PLATE 41. *Heterorotula lucasi*: Gemmular theca. **A**, Gemmule having multilayer theca armed by radial gemmuloscleres around the central cavity bearing a mass of totipotent cells; **B-D**, trilayered theca with spiny birotules embedded in the chambered pneumatic layer, proximal rotules of gemmules adherent to the multilayer inner membrane, central cavity with totipotent cells.

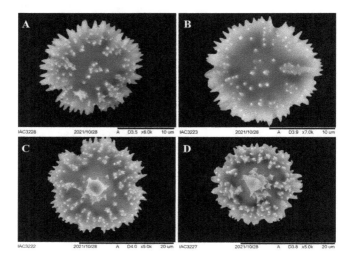

PLATE 42. *Heterorotula lucasi*: Rotule dorsal and ventral. **A-B**, Dorsal view showing crenulated margins; **C-D**, ventral view showing crenulated margins.

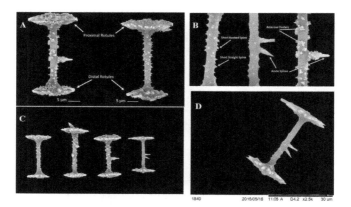

PLATE 43. *Heterorotula lucasi*: Birotulate gemmuloscleres. **A**, Difference in size of proximal and distal rotules; **B**, types of spines found on gemmuloscleres; **C**, variation in gemmulosclere size; **D**, gemmulosclere having two acute spines, scale bar for size determination.

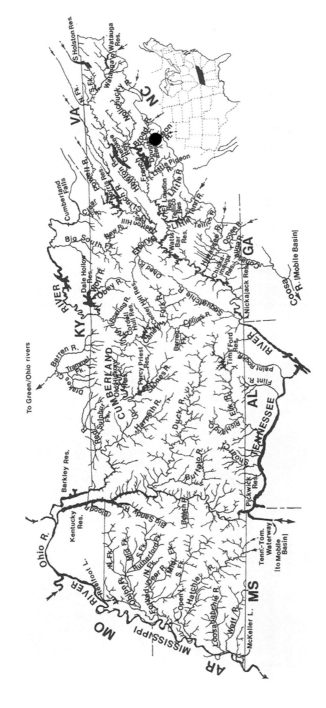

Heterorotula lucasi.

supports the dermal membrane. **Choanosomal skeleton** as a network of multi-spicular fibers, with scanty spongin. **Basal spongin** place well-developed. **Megascleres** fusiform acanthoxeas 223.1–335 μm (276.8 ± 24.7) in length and 7.7–13.7 μm (10.9 ± 1.0) in width, slightly curved, with variable dense spines except toward the variable pointed tips, to less frequently nearly spineless. Acanthoxeas shaft from slender in ectosomal area to stouter in the endosome. **Microscleres** absent. **Gemmules** scattered in **skeletal** network, subspherical, 448–613 μm (528 ± 55.9) in diameter. Foramen simple with smooth undulated margins, slightly elevated above gemmular surface. **Gemmular cage** of acanthoxeas with small spines. **Gemmular theca** trilayered ≈ 50 μm in thickness. **Outer layer** fibrous to compact with distal rotules more or less embedded. **Pneumatic layer** well-developed and thick, ranging in the same gemmule from mainly fibrous to chambered network of irregularly polygonal meshes of variable size. **Inner layer** multilayered of 3–5 layers of compact spongin. **Gemmuloscleres** radially embedded as a dense monolayer in pneumatic layers of theca, with distal smaller rotules covered by outer layer and proximal larger

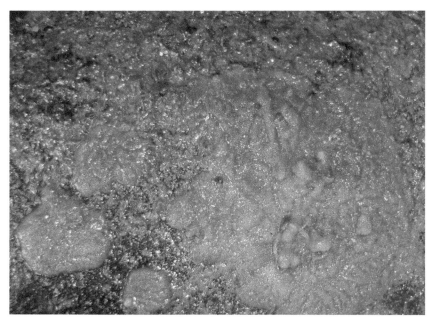

FIG. 39. *Heterorotula lucasi*

rotules, partly overlapping one to another, not embedded into inner layer. **Gemmuloscleres** slender birotules 19.8–48.6 μm (35.1 ± 5.5) in length, with narrow spiney shaft 2.7–4.4 μm (3.3 ± 0.3) in width. Shaft spines of 3 types (a) simple, short curved to straight, smooth spines, (b) spines with tips arranged in asterose clusters (microspines in rosettes), and (c) large, acute spines up to 3 μm long bearing secondary microspines. **Rotules** flat with crenulated notched to shallowly incised margins and both rotules inner and outer surfaces bearing numerous microspines sometimes in radial rows of significantly different diameters. Large proximal rotules 19.4–24.4 μm (21.6 ±1.1) in diameter, small distal rotules 16.6–21.7 μm (18.9 ±1.1) in diameter.

Heterorotula lucasi is currently endemic to Tennessee.

RACEKIELA RYDERI (POTTS, 1880)

Species Description: **Color** in vivo: green due to green alga symbiont. **Consistency** in vivo and when dried soft with loose texture. **Growth form** (Fig. 40) massive, cushion to dome–like, with lobes. **Surface** lobed. **Oscula** with five–six radial canals. **Ostia** scattered. **Ectosomal skeleton** without special architecture. **Choanosomal skeleton** irregular, fragile network of paucispicular ascending fibers joined by secondaries. **Spongin** scanty. **Megascleres** are oxeas 197–297 (246 ± 20) μm in length, with conical spines, which show a great deal of variation from habitat to habitat. **Microscleres** are absent. Larger gemmuloscleres 44.4–64.1 (57 ± 4.8) μm in length, have recurved spines on their shaft and recurved hooks 13–17.8 (15.3 ± 1.2) μm in diameter on their ends. **Gemmules** 320–372 (343 ± 18) μm in diameter, spherical. **Foramen** simple,

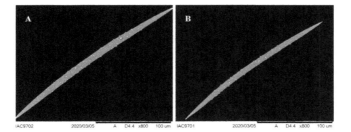

PLATE 44. *Racekiela ryderi*: Megascleres. **A-B**, Sparsely spined oxeas, spines absent near tips of spicules; **C-D**, oxeas, nontypical, completely spined to tips.

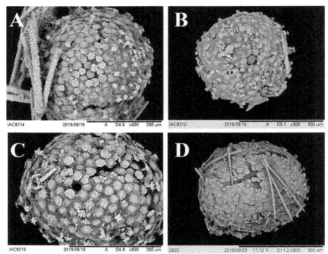

PLATE 45. *Racekiela ryderi* Gemmules. **A**, Nontypical, completely spined megascleres on left side of gemmule; **B**, heavily armed with gemmulosclercs, some protruding through the outer membrane, closed foramen; **C**, open foramen, rotules of gemmulosclercs; **D**, gemmule with typical megascleres on its surface.

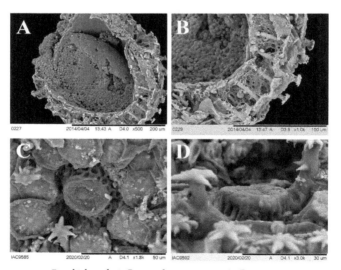

PLATE 46. *Racekiela ryderi*: Gemmular structures. **A**, Cross section of gemmular theca showing staminal cells; **B**, close-up of trilayered gemmular theca, (from inside out) inner membrane of spongin, gemmulosclercs embedded in pneumatic layer, outer spongin layer with protruding gemmulosclercs; **C-D**, short simple foramina.

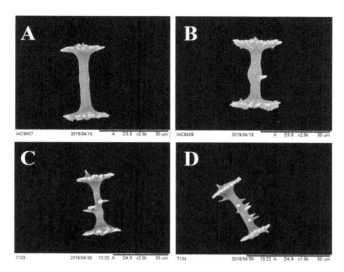

PLATE 47. *Racekiela ryderi*: Gemmuloscleres. Shorter type, **A-D**, birotules having flattened rotules with numerous small teeth.

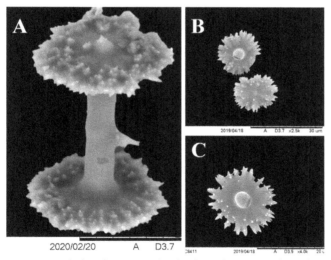

PLATE 48. *Racekiela ryderi*: Gemmulosclere birotules. **A**, Spiny birotule with spine on shaft; **B**, dorsal and ventral views of spiny-toothed rotules; **C**, ventral view of rotule.

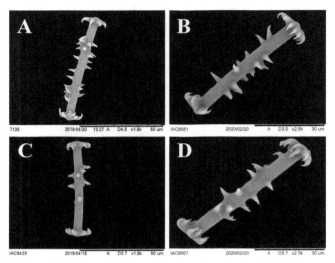

PLATE 49. *Racekiela ryderi:* Gemmuloscleres, larger type. **A-D**, Shafts with recurved spines and tips having pseudorotules of recurved hooks.

FIG. 40. *Racekiela ryderi*

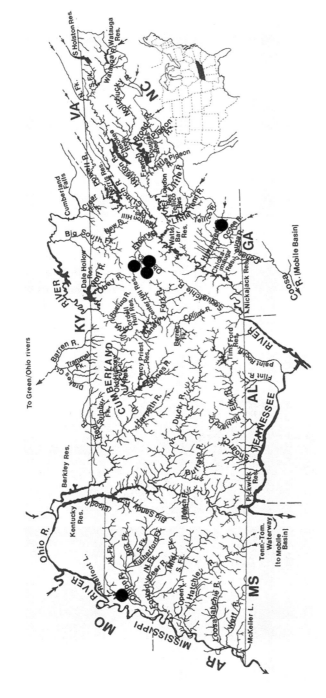

Racekiela ryderi.

short on a conical elevation. **Gemmular theca** is trilayered. **The outer layer** of compact spongin with protruding gemmuloscleres. **Pneumatic layer** spongin having small round (more or less) chambers with gemmuloscleres radially arranged with the longer group protruding from the gemmular surface. **Inner layer** of compact spongin with sublayers. **Gemmuloscleres** are birotules of two distinct length groups. Shorter gemmuloscleres 28.9–45.4 (35.5 ± 5) μm in length, shafts armed with one or a few spines and flattened rotules 17.3–24.8 (20.5 ± 1.6) μm in diameter, having numerous small teeth.

In Tennessee *R. ryderi* is common in the Emory River drainage. It's known from the Nearctic, Palearctic, and Neotropical Regions.

RADIOSPONGILLA CEREBELLATA (BOWERBANK, 1863)

Species Description: **Color** in vivo: yellowish, green due to green alga symbiont. **Consistency** in vivo soft with loose texture. **Growth form** encrusting. **Oscula** conspicuous. **Ostia** scattered. **Ectosomal skeleton** with immature gemmuloscleres present in dermal membrane, no special architecture. **Choanosomal skeleton** an irregular network having ascending paucispicular

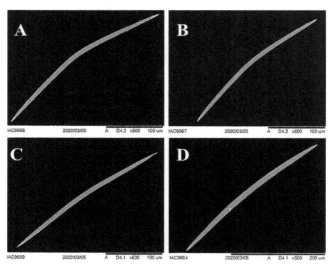

PLATE 50. *Radiospongilla cerebellata*: Megascleres. **A-D,** Straight to slightly curved smooth oxeas.

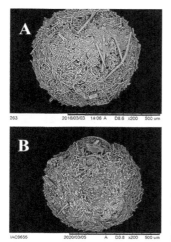

PLATE 51. *Radiospongilla cerebellata*: Gemmules. **A-B**, Outer layer of compact spongin with gemmuloscleres on surface.

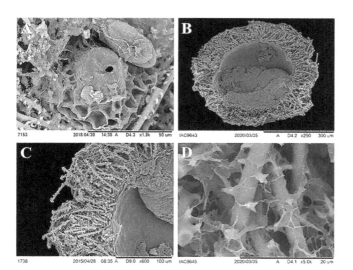

PLATE 52. *Radiospongilla cerebellata*: Gemmular structures. **A**, Tubular foramen; **B**, (from inside-out) staminal cells, inner membrane, gemmuloscleres, outer membrane of compact spongin; **C**, Gemmular theca with tangential and radial arrangement of gemmuloscleres; **D**, pneumatic layer embedded with gemmuloscleres.

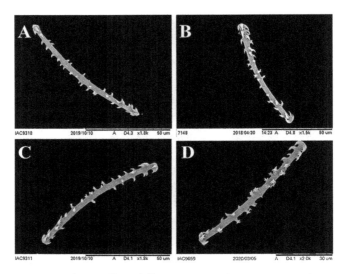

PLATE 53. *Radiospongilla cerebellata*: Gemmuloscleres. **A-D**, Curved, rarely straight strongyles, covered with numerous spines, recurved spines at tips.

primary fibers joined by secondaries. **Megasclere** essentially smooth, straight, sharp pointed oxeas 194–385 (307.3 ± 47.6) μm in length. **Microscleres** are absent but immature gemmuloscleres may be present in some portions of the dermal membrane. **Gemmules** 451–583 (506 ± 49) μm in diameter, subspherical, single, and scattered. **Foramen** slender straight tube. **Gemmular theca** trilayered with gemmuloscleres arranged in two layers tangentially on the outer layer and radially on the inner. **Outer layer** of compact spongin. The **Pneumatic layer** of fibrous spongin poorly developed. **Inner layer** of compact spongin with sublayers. Gemmuloscleres are embedded in the pneumatic layer and arranged into two layers with those in the inner layer arranged radially and those in the outer tangentially. **Gemmuloscleres** are curved, rarely straight, strongyles 65–94 (78.3 ± 6.6) μm in length, covered with numerous spines recurving toward terminal ends.

Some confusion exists surrounding this sponge occurring in the United States. The World Porifera Database (van Soest, et al., 2017) distribution map for *R. cerebellata* does not show this species occurring in the western

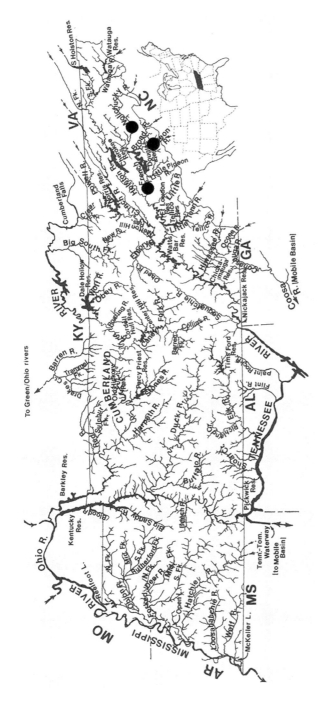

Radiospongilla cerebellata.

hemisphere. However, it has been reported from Texas (Poirrier, 1972) in the Cahaba River and Shades Creek in Alabama (Mobley, 2010), eastern Tennessee by Kunigelis and Copeland (2014), and a fish hatchery in Louisiana (Michael Poirrier, personal communication). In Tennessee *R. cerebellata* has been collected from eastern headwater rivers of the Tennessee River drainage. Phenotypically specimens from the United States appear to be *R. cerebellata*. Molecular studies comparing DNA from known *R. cerebellata* specimens to DNA of United States specimens are needed to rectify this situation.

RADIOSPONGILLA CRATERIFORMIS (POTTS, 1882)

Species Description: **Color** in vivo: flesh color, green due to green-algae symbiont. **Consistency** in vivo soft. **Growth form** cushion. **Surface** is even. **Oscula** not conspicuous but numerous. **Ostia** scattered. **Ectosomal skeleton** without any special architecture. **Choanosomal skeleton** irregular ascending paucispicular primary fibers joined by secondaries. **Spongin** scanty. **Megascleres** are slender, sharply pointed, sparsely microspined except at tips oxeas 207–274 (240.8 ± 16.1) μm in length. **Microscleres** are absent. **Gemmules** 347–462

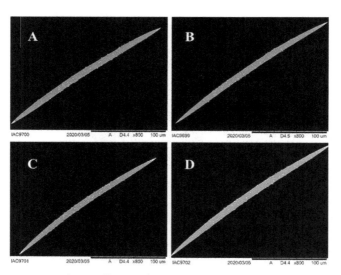

PLATE 54. *Radiospongilla crateriformis*: Megascleres. **A-D**, Oxeas, slender, slightly curved, except for tips sparsely covered by minute spines.

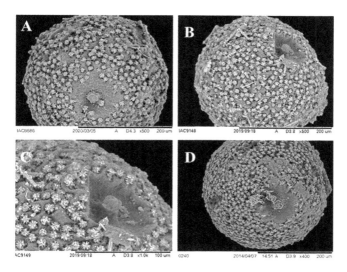

PLATE 55. *Radiospongilla crateriformis*: Gemmules. **A-D**, Crater-like depressions with short tubular foramenia, gemmuloscleres protruding through outer spongin membrane.

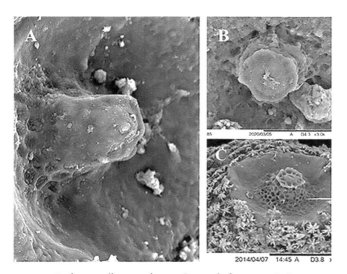

PLATE 56. *Radiospongilla crateriformis*: Gemmule foramina. **A-C**, Close-up views of foramina within crater-like depressions.

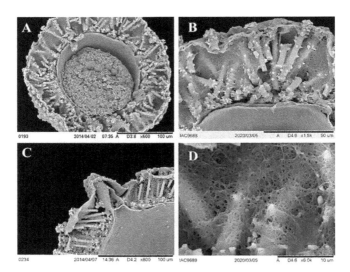

PLATE 57. *Radiospongilla crateriformis*: Gemmular theca. **A**, Cross section, (from inside out), staminal cells inner membrane, radially arranged single layer of gemmuloscleres, pneumatic layer, outer membrane; **B**, inner membrane of compact spongin, gemmuloscleres within pneumatic layer; **C**, cross section of gemmular theca through foramen; gemmuloscleres extend from inner membrane to protruding the outer membrane; **D**, close-up of gemmuloscleres within pneumatic layer.

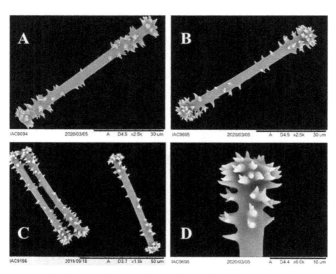

PLATE 58. *Radiospongilla crateriformis*: Gemmuloscleres. **A-C**, Slender, spiny, strongyles, clusters of recurved spines at tips forming pseudorotules; **D**, close-up of pseudorotule.

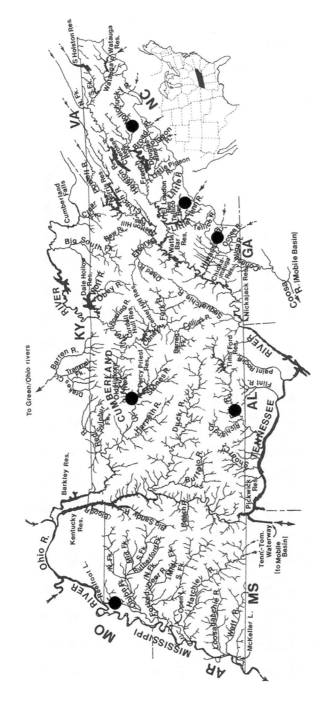

Radiospongilla crateriformis.

(404 ± 39) μm in diameter, subspherical occur as singles and scattered. **Foramen** is a short tube not reaching the level of the outer membrane. **Gemmular theca** trilayered. The **Outer layer** sometimes present, when present armed by distal ends of gemmuloscleres. **Pneumatic layer** is a dense network of spongin fibers forming small irregular meshes, with monolayer of radially embedded gemmuloscleres. **Inner layer** of compact spongin with sublayers. **Gemmuloscleres** are pseudobirotules 52–77 (64.7 ± 5.7) μm in length, slender, shafts with small conical spines only at ends or all over but always more abundant at ends, terminal ends comprised of several rows of recurved spines 7.6–14.4 (11 ± 1.4) μm in diameter. Gemmuloscleres in the area surrounding the foramen are arranged in an inward slanting manner which results in a crater-like depression.

Radiospongilla crateriformis is reported from the eastern half of the United States with reports from Texas and Wisconsin. Also, reported from Canada, China, Japan, Southeast Asia, and Australia.

SPONGILLA LACUSTRIS (LINNAEUS, 1758)

Species Description: **Color** in vivo: yellowish to sandy brown, green due to green alga symbiont. **Consistency** in vivo soft and fragile. **Growth form** is variable from encrusting to massive, can be branched, or arborescent up to 40–50 cm in height. **Surface** hispid. **Ectosomal skeleton** no special architecture. **Choanosomal Skeleton** an irregular network from isotropic in encrusting sections to anisotropic in branches; primary fibers pauci– to multi–spicular joined by paucispicular secondary fibers. **Spongin** is abundant. **Megascleres** slightly curved to straight entirely smooth oxeas 165–280 (213.2 ± 25.1) μm in length. **Microscleres** are slightly curved oxeas 41.4–79.9 (59.5 ± 8.4) μm in length, completely covered with small spines. **Gemmules** 343–447 (384 ± 35) μm in diameter, subspherical to suboval, in clusters or scattered. Two forms of gemmules can be found unarmored (those lacking gemmuloscleres) and armored (those having gemmuloscleres); may find a gemmular cage composed of megascleres around gemmules lacking gemmuloscleres. **Gemmular theca** of two forms: may be thick walled (trilayered) which have well-developed outer, pneumatic, and inner layers or may be thin walled which are monolayered with a variable number of sublayers which correspond to the inner layer of thick walled gemmules. **Foramen** slightly elevated, may or may not have a collar. Gemmules may have multiple foramina (up to six). Gemmuloscleres are present in thick walled gemmules, usually absent in thin walled gemmules.

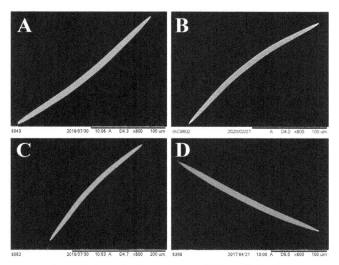

PLATE 59. *Spongilla lacustris*: Megascleres. **A-D**, Smooth, fusiform, slightly curve too nearly straight oxeas.

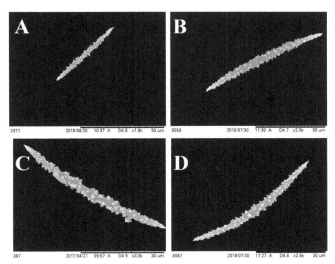

PLATE 60. *Spongilla lacustris*: Microscleres. **A-D**, Straight to slightly curved fusiform oxeas bearing dense spines along entire length of spicule, some spines with microspines creating a rosette appearance.

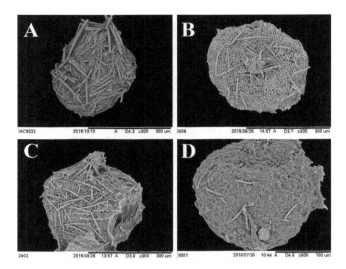

PLATE 61. *Spongilla lacustris*: Gemmules. **A**, Upper region showing remains of a gemmular cage composed of megascleres; **B-C**, areas with missing thin outer spongin membrane exposing honeycomb-like design of pneumatic layer, elevated foramina, and scattered microscleres; **D**, outer spongin layer with microscleres.

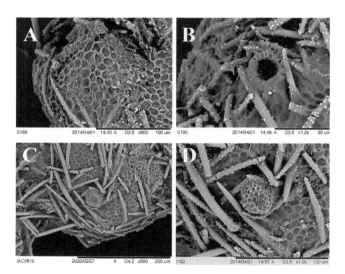

PLATE 62. *Spongilla lacustris*: Gemmule surface. **A**, Close-up view of outer spongin membrane; **B** close-up view of open foramen; **C-D**, two views of the same closed foramen, megascleres and microscleres embedded in and at surface of outer spongin layer.

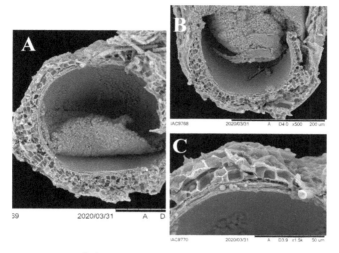

PLATE 63. *Spongilla lacustris*: Gemmular thecae. **A-B**, Thick-walled, unarmored (no gemmuloscleres); (from inside out) staminal cells, inner membrane, well-developed pneumatic membrane, outer membrane; **C**, close-up of pneumatic membrane and multilayers of inner membrane.

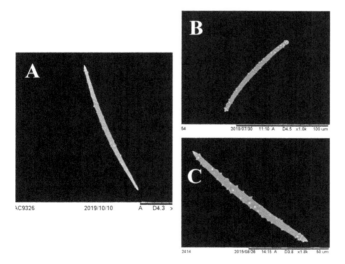

PLATE 64. *Spongilla lacustris*: Gemmuloscleres. **A,** Slightly bent, slender, sparsely spined oxea; **B**, strongyle form; **C**, nearly straight, spined oxea.

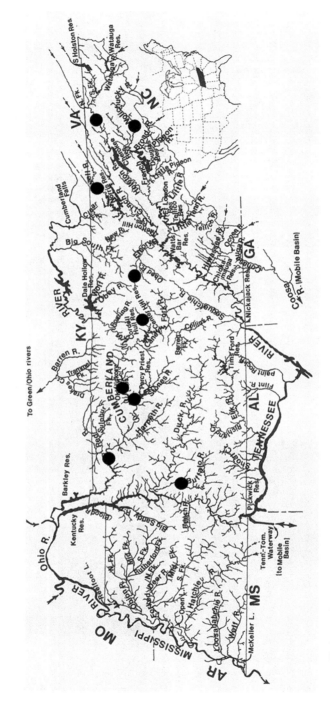

Spongilla lacustris.

Gemmuloscleres slightly to strongly curved oxeas or strongyles having strong, curved spines concentrated more toward tips of spicules. In Tennessee specimens gemmuloscleres measured 66–142 (97.9 ± 12.9) μm in length.

Distribution restricted to the Northern Hemisphere. Reported from Europe, Russia, China, Japan, Canada, and United States. In Tennessee *S. lacustris* is broadly distributed.

TROCHOSPONGILLA HORRIDA (WELTNER, 1893)

Species Description: **Color** in vivo light yellow to dark brown. **Consistency** in vivo firm to moderately hard. **Growth form** (Fig. 41) encrusting. **Surface** hispid. **Oscula** scattered. **Ostia** scattered. **Ectosomal skeleton** no special architecture. **Choanosomal skeleton** anisotropic with pauci- to multispicular fibers joined by secondaries. **Spongin** scanty. **Megascleres** straight to slightly curved oxeas 145–210 (188.4 ± 12.6) μm in length, pointed, and covered with stout spines. **Microscleres** are absent. **Gemmules** 314–372 (344 ± 19) μm in diameter, subspherical may be found single or in carpets at base of sponge; enveloped in a multigemmular pneumatic pseudo-cage (not the one in the gemmular theca) of spongin trabecula's armed by megascleres. **Foramen** conical

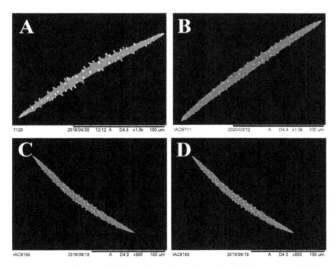

PLATE 65. *Trochospongilla horrida*: Megascleres. **A-D**, Straight to slightly curved oxeas, covered by stout, blunt, truncated spines.

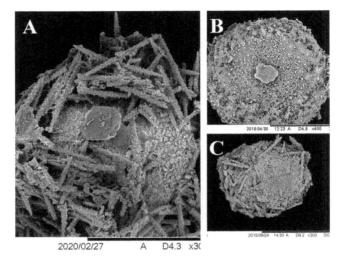

PLATE 66. *Trochospongilla horrida*: Gemmules. **A**, Covered with megascleres, conical foramen, protruding rotules of gemmuloscleres at surface, scant surface spongin; **B**, closed conical foramen with collar; **C**, rotules of gemmuloscleres cover by thin layer of spongin.

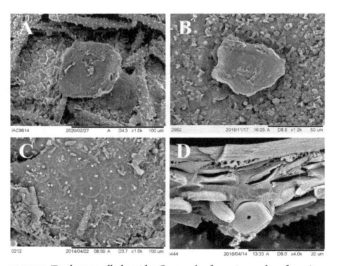

PLATE 67. *Trochospongilla horrida*: Gemmular foramina and surface. **A** and **B**, Close-up of foramina; **C**, close-up of surface rotules with thin layer of spongin; **D**, cross section of gemmular theca showing layers of inner spongin membrane (at top), reinforced with overlapping rotules of adjacent gemmuloscleres, thin outer membrane.

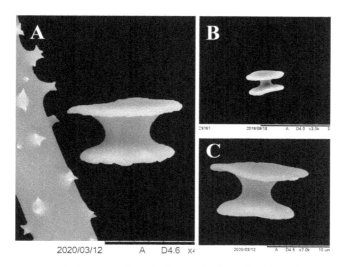

PLATE 68. *Trochospongilla horrida*: Gemmuloscleres. **A-C**, Birotules having small, stout, smooth shaft; **A**, notice megasclere spines can have 2 to 3 small, sharp barbs.

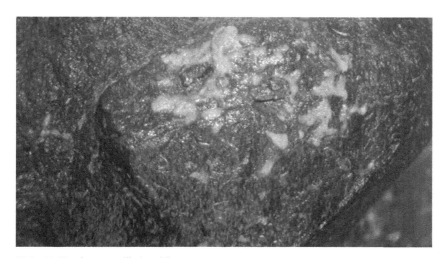

FIG. 41. *Trochospongilla horrida*

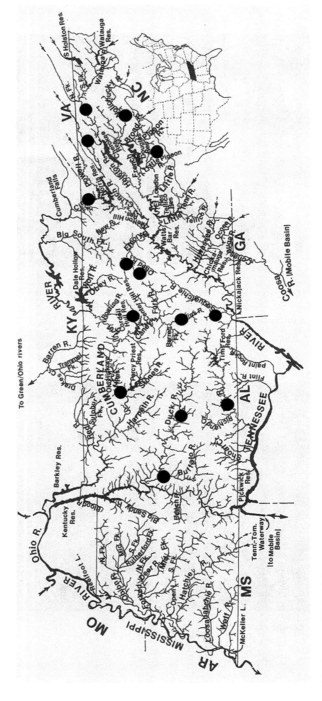

Trochospongilla horrida.

or tubular with collar. **Gemmular theca** is trilayered **outer layer** of compact spongin with sublayers, reinforced by one of two layers of embedded gemmuloscleres having overlapping proximal rotules. **Pneumatic layer** present. **Inner layer** of spongin sublayered. **Gemmuloscleres** minute birotules 6.6–9.5 (8 ± 0.68) μm in length, having smooth shafts. Rotules of gemmuloscleres with more or less recurved margins of slightly different diameters, smaller rotules 10.3–13.2 (11.8 ± 0.6) μm in diameter and larger rotules 12.3–15.6 (14.5 ± 0.7) μm in diameter. Overall length of gemmuloscleres is never greater than the diameter of the smaller rotule.

Trochospongilla horrida is widely distributed in the Northern Hemisphere. Reported from Canada, Central America, China, Europe, and United States. A difference exists between European and American specimens. European specimens have a distinct class of oxeas similar to megascleres which serve as a special dermal membrane spicule and is also used for forming the outer capsule of gemmules. This characteristic has not been documented for American specimens.

TROCHOSPONGILLA LEIDYI (BOWERBANK, 1863)

Species Description: **Color** in vivo: light gray to drab. **Consistency** in vivo hard and firm. **Growth form** encrusting. **Ectosomal skeleton** no special architecture. **Choanosomal skeleton** regular network of paucispicular fibers and secondary fibers. **Spongin** is scanty except for that of the gemmular theca. **Megascleres** in most of the literature oxeas are smooth, Tennessee specimens straight to slightly curved oxeas covered with minute spines 105–166 (138 ± 12.3) μm in length. **Microscleres** are absent. **Gemmules** 358–453 (408 ± 28) μm in diameter, subspherical. **Foramen** short, tubular, or conical. **Gemmular theca** trilayered surrounded by a cage of megascleres. **Outer layer** of compact spongin. **Pneumatic layer** a thin layer of small rounded meshes, with a monolayer of radially embedded gemmuloscleres. **Inner Layer** compact spongin having sublayers. **Gemmuloscleres** are minute birotules 8.1–11.7 (9.7 ± 0.8) μm in length having a stout, short shaft. Rotules of gemmuloscleres are circular 13.1–17.8 (15.1 ± 1) μm in diameter, with recurved margins.

In Tennessee *T. leidyi* has been collected only from the Cumberland River drainage within the Nashville Basin. Two spellings of the specific epithet for this species can be found in the literature. Bowerbank (1863) used both *leidyi* and *leidii* in his description of this species. In plate XXXVIII of his publication, which shows drawings of spicules, Bowerbank uses *leidyi* but in the narrative

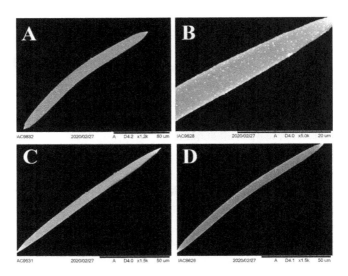

PLATE 69. *Trochospongilla leidyi*: Megascleres. **A-D**, Straight to slightly curved oxeas; **A** and **C** appear to be smooth but under greater magnification can be seen to be microspined as shown in **B** and **D**.

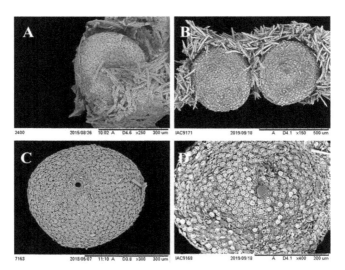

PLATE 70. *Trochospongilla leidyi*: Gemmules. **A-B**, Gemmules within cradle of megascleres; **C**, rotules of gemmuloscleres on gemmule surface, open foramen; **D**, closed foramen, gemmulosclere rotules above thin outer membrane.

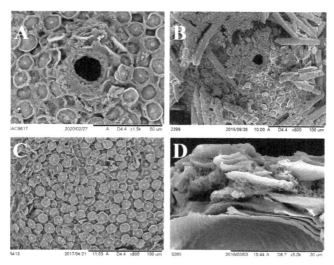

PLATE 71. *Trochospongilla leidyi*: Gemmule surface and theca. **A**, Close-up of gemmule surface showing overlapping rotules of adjacent gemmuloscleres and open foramen; **B**, cradle of megascleres, foramen aperture; **C**, overlapping rotules of adjacent gemmuloscleres; **D**, gemmular theca (from inside out), multilayered inner membrane, gemmuloscleres embedded within pneumatic layer, thin outer membrane of spongin.

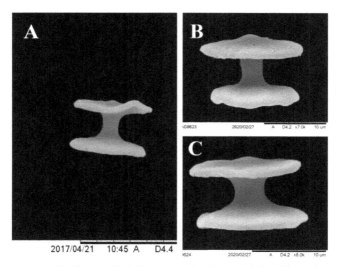

PLATE 72. *Trochospongilla leidyi*: Gemmuloscleres. **A-C**, Small birotules having circular rotules of equal size.

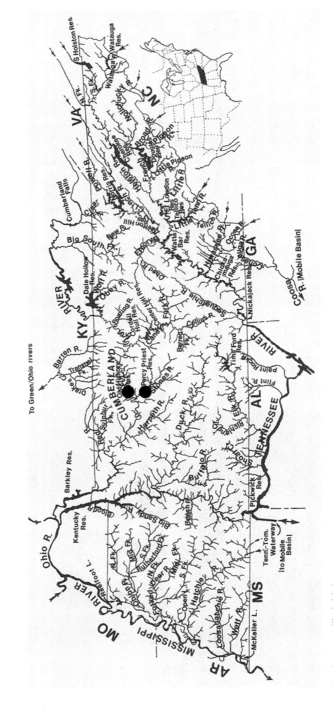

Trochospongilla leidyi.

describing this species he uses *leidii*. We agree with Manconi and Pronzato (2016) that the correct specific epithet should be *leidyi* because the name etymology refers to the American paleontologist Joseph Leidy (1823–1890). From outside the United States *T. leidyi* has been reported from the Gatún Locks, of the Panama Canal (Jones and Rützler 1975).

The megascleres of *T. leidyi* are reported in much of the scientific literature to be entirely smooth. However, Jones and Rützler (1975) reported megascleres of *T. leidyi* found in the Panama Canal to appear smooth under the light microscope but when viewed under a SEM to be covered by minute spines about 0.1 μm tall. Scanning electron microscopy of megascleres of Tennessee specimens revealed they also have minute spines.

Megascleres for Tennessee specimens appear to be much shorter than the 150–170 μm lengths reported by Penney and Racek (1968), Reiswig, et al. (2010) and Manconi and Pronzato (2016). While most researchers have treated *T. leidyi* megascleres as representing a single size group, Jones and Rützler (1975) recognized two groups of megasclere oxeas in their Panama Canal specimens.

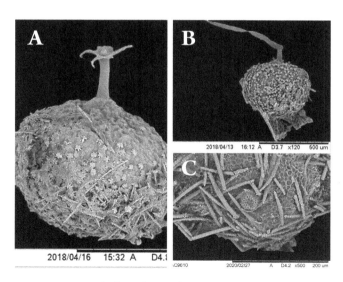

FIG. 42. Gemmule foramina. **A**, foramen with several short cirri; **B**, foramen with a single long cirrus; **C**, foramen without a cirrus.

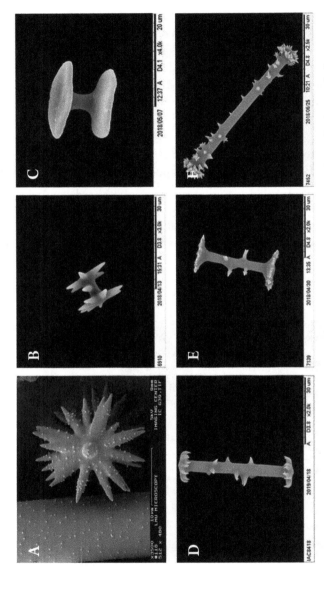

FIG. 43. Examples of gemmuloscleres. **A,** with incised rotules; **B,** with incised rotules and spine on shaft; **C,** with smooth rotules and shaft; **D,** rotules of hooks and spiny shaft; **E,** flat rotules and spiny shaft; **F,** pseudorotules of recurved spines

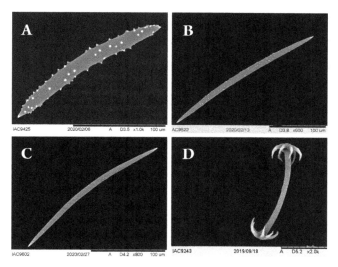

FIG. 44. Megascleres and a single type of microsclere. **A**, spiny oxea; **B**, spiny oxea; **C**, smooth oxea; **D**, microsclere with micropseudobirotules of hooks.

They reported oxeas of the main skeleton to have a length range of 125–175 μm with an average of 149 μm and oxeas of the gemmular capsule to have a range of 105–155 with an average of 141.9 μm. Even though megascleres of Tennessee specimens were treated as a single group, their average length measurement of 138 μm is similar to those of Panama Canal specimens than those collected elsewhere.

Dichotomous Key to Spongillida of Tennessee

The following dichotomous key is limited to the species discussed in this book and is based on characteristics of spicules and gemmules. To successfully identify a sponge to species, all spicule types and gemmules of that species must be present and carefully viewed. If gemmules are present but gemmuloscleres are absent the species is more than likely *Spongilla lacustris* because this species produces winter gemmules which lack gemmuloscleres. Figures 42–44 provide SEM microphotographs helpful in the use of the following key.

Dichotomous Key for the Freshwater Sponges of Tennessee

1a. Gemmuloscleres are "c," bean, or button shaped—*Cherokeesia armata*
1b. Gemmuloscleres are not c, bean or button shaped → 2
2a. Microscleres present → 3
2b. Microscleres absent → 6
3a. Microscleres micropseudobirotules of hooks—*Corvospongilla becki*
3b. Microscleres oxeas or strongyles → 4
4a. Megascleres smooth, foramen of gemmule without cirrus—*Spongilla lacustris*
4b. All or some megascleres with spines, foramen of gemmule with distinct, terminal cirrus → 5
5a. Foramen of gemmule with a single long cirrus (occasionally 2)— *Heteromeyenia latitenta*
5b. Foramen of gemmule with several cirri projections—*Heteromeyenia tubisperma*
6a. Megascleres smooth or with minute (approximately 0.1 μm tall) spines → 7
6b. Megascleres all or some with spines greater than 0.1 μm in length → 10
7a. Gemmuloscleres are of two types: curved strongyles covered with spines and usually a few sparsely spined oxeas—*Eunapius fragilis*
7b. Gemmuloscleres birotules or terminal ends with hooks → 8
8a. Terminal ends and shafts of gemmuloscleres with hooks— *Radiospongilla cerebellata*
8b. Gemmuloscleres birotules → 9
9a. Gemmuloscleres birotules without incised teeth, shaft completely smooth, rotule margins recurved—*Trochospongilla leidyi*
9b. Gemmuloscleres birotules with incised teeth, shaft smooth or with a few spines—*Ephydatia fluviatilis*
10a. Gemmuloscleres birotules without incised teeth, shaft completely smooth—*Trochospongilla horrida*
10b. Gemmuloscleres birotules with incised teeth or pseudobirotules with spines → 11
11a. Gemmuloscleres pseudorotules composed of recurved spines— *Radiospongilla crateriformis*
11b. Gemmuloscleres birotules →12
12a. Proximal and distal rotules of birotules of different diameters— *Heterorotula lucasi*
12b. Proximal and distal rotules of birotules of essentially same diameter → 13

13a. Gemmulosclere birotules of two distinct length groups, shorter forms with flattened rotules and a few teeth on shaft, larger forms numerous recurved spines on shaft and recurved hooks on the ends of shaft—
Racekiela ryderi
13b. Gemmuloscleres are birotules of one size with deeply incised teeth—
Ephydatia muelleri

CHAPTER 7

Environmental Variables and Organismal Interactions

> "Every man is but a spunge,
> and but a spunge filled with teares..."
> —JOHN DONNE (1630)—

At their broadest classification, bodies of water are classified as lotic and lentic waters. Lotic, or riverine habitats, are flowing waters such as springs, streams, and rivers. Streams and rivers are classified as montane, upland, or lowland. Montane waters are high gradient, fast flowing, with rocky substrate. Upland waters have a moderate gradient and flow with primarily rocky substrate. Lowland waters are slow moving, low gradient waters with substrates of sand and silt. A lentic ecosystem is a body of still water. Vernal pools, swamps, ponds, and lakes are classes of lentic ecosystems.

Many different chemical and physical abiotic factors determine where a species can survive. Every species has an optimum range for each abiotic factor needed for its survival. Unfortunately, there is no habitat that provides the optimum range for every abiotic factor needed by a species. This is why biologists are concerned with a species range of tolerance for abiotic factors. The range of tolerance is the range of values from the minimum needed to the maximum tolerated by a species. A biological hypothesis known as the niche breath hypothesis assumes a species cannot exist in environments where values of an important abiotic factor fall below the minimum or above the maximum. Furthermore, this hypothesis assumes species having broad tolerances to abiotic factors occupy larger geographic areas because they have greater intrapopulation

Hiwassee River, Polk County, Tennessee.

genetic variation than those having narrow tolerances. Genetic variation typically results in more phenotypic variation, which allows some individuals of a population to adapt to changing environments. Why? Because natural selection acts only on phenotypes.

Abiotic factors known to influence sponge populations include pH, silica, substrate availability, and sunlight for endosymbiotic algae photosynthesis.

The hydrogen ion concentration (pH) of water determines the solubility of compounds and nutrients and their availability to aquatic life. A pH range of 6.5–8 is considered optimum for many aquatic organisms. When the pH of a system changes nutrients can become unavailable to living organisms.

Phosphorus is affected by changes in pH. At alkaline pH values, phosphate ions react with calcium and magnesium to form less soluble compounds. At acidic pH values, phosphate ions react with aluminum and iron again forming less soluble compounds. Most nutrients tend to become less available as pH becomes more acidic. Also, as waters become more acidic heavy metal toxicity increases.

Sponge reproduction can be reduced or prevented at low pH. As an example, consider the results of a study of the effects of low pH on gemmule hatching rate and hatchability. Gemmules collected from three populations of *Ephydatia muelleri* were placed in waters of low pH (6.5–5.8) at 5° C for one week. A differential but lower hatching rate and hatchability was found for all three gemmule populations. When placed in waters with a pH of 5, gemmule hatching ceased.

Freshwater sponges need silica to produce spicules. Some species are restricted to waters having high silica levels, other species to low levels. Populations of *Spongilla lacustris* are distributed across habitats having significant differences in silica levels. In Wisconsin, lakes having low silica levels specimens of *Spongilla lacustris* produced weaker, easier to break spicules than specimens living in high silica lakes.

The quantity and availability of suitable substrates determines population density and individual sponge growth. The absence of hard substrate is a limiting factor. Its absence completely prevents the presence of most freshwater sponge species. Only *Spongilla lacustris* is known to routinely live on soft substrates. The permanency of substrates is important. Individual sponges attached to large boulders, logs, or to human-made objects such as bridge supports can maintain their existence through repeated cycles of active sponge to gemmule to active sponge for long periods of time. Anthropogenic structures such as bridges, weirs, or dams composed of concrete, brick, natural stone, or wood provide additional sites for colonization and can be especially important in areas of sand, silt, and gravel substrates.

As previously stated, sponge growth is enhanced by the presence of symbiotic green algae. Light penetration is one of the factors influencing the productivity of algae and in turn the availability of glucose provided to a sponge. The availability of plant-produced glucose affects not only individual sponge growth, but also increases reproductive potential.

The following information provides ranges of environmental variables which influence growth, population density, and geographic distributions of freshwater sponges. Range values of abiotic variables were derived by combing the results of several scientific reports, except where noted. Ecological interactions with other organisms and habitat information are included. For some species, data is lacking. See Tables B1 and B2 in the appendix for distribution data by state and by freshwater sponge species.

Cherokeesia armata

Habitat: lotic waters
Substrate: rocks
Associations with other life-forms: *Eunapius fragilis* was collected with *Cherokeesia armata* at every *C. armata* collection site.
Ranges of abiotic variables: The following data is from 24 samples taken at a *C. armata* collection site on Nolichucky River, Tennessee.
  DO_2: 4.3–8.8 parts per million (ppm)
  Nitrates: 0.52–0.79 ppm
  Nitrites: 0.001–0.60 ppm
  pH: 7.0–8.0
  Phosphates: 0.06–1.95 ppm
  SiO_2: 2–11 ppm
  Sulfates: 2–35 ppm
  Total Ammonia: .89–2.06 ppm

Corvospongilla becki

Habitat: Lentic in coastal waters of Louisiana, some of which are slightly brackish, lotic waters in Alabama and Tennessee
Substrate: rocks in lotic waters
Associations with other life-forms: In Louisiana *Corvospongilla becki* was associated with the bryozoans *Plumatella repens* (Linnaeus, 1758), *Pottsiella erecta* (Potts, 1887), and the entoproct *Urnatella gracile* Leidy, 1851, and the brackish water amphipod, *Corophium lacustre* Vanhöffen, 1911. Sponges collected from the same Louisiana lake were *Eunapius fragilis*, *Trochospongilla horrida*, and *Trochospongilla leidyi*. In Tennessee *Trochospongilla horrida* was collected at the same site as *Corvospongilla becki*.
Values of abiotic variables: The following data values are from a one-time sample at collection site on Duck Lake, Louisiana.
  Chloride: 95 ppm
  Conductivity: 614 micromho/cm (μmhos/cm)

Free CO_2: 22 ppm
pH: 7.1
Total Alkalinity: 125 ppm

Ephydatia fluviatilis

Habitat: lotic and lentic, tolerant of brackish water, alkaline waters rich in calcium, salt lakes, and caves. In Louisiana *Ephydatia fluviatilis* was found to survive periods of high siltation. In Michigan this species was reported to tolerate polluted waters. Known to colonize and have high abundance in habitats that dried out on a regular basis.
Substrate: include rocks, logs, shells of freshwater mussels, man–made materials (plastic, concrete, metal, and glass objects).
Associations with other life-forms: Parasitized by the spongillafly *Sisyra vicaria* (Walker, 1953). Associated with the invasive zebra mussel, *Dreissena polymorpha*. In the UK *Ephydatia fluviatilis* occurred frequently with *Spongilla lacustris* and *Eunapius fragilis* at canal sites. Known to have algae symbionts. Serve as a host for bacteria of the genus *Pseudomonas* which have antimicrobial activities.
Ranges of abiotic variables:
    Alkalinity: methyl orange: 150–230 ppm $CaCO_3$
    Calcium: 68–82ppm
    Conductivity: 9–921.67 μmhos/cm
    DO_2: 1–14.73 ppm
    Free CO_2: 6.5–9 ppm
    Magnesium: 8–28 ppm
    Oxidation Reduction Potential: 79–921.67 mV
    pH: 5.9–8.48
    Salinity: 0.05–0.57 practical salinity unit (psu)
    Temperature: 7.86–33° C
    Total Dissolved Solids: 0.07–0.74 ppm
    Total Hardness: 140–180 ppm $CaCO_3$

Ephydatia fluviatilis is known to be sensitive to low concentrations of heavy metals. Exposure to 0.001mg/l of cadmium or mercury resulted in gemmule malformations.

A positive association between the growth of *Ephydatia fluviatilis* and waters having high alkalinity levels has been reported.

Ephydatia muelleri

Habitat: lotic and lentic waters
Substrate: rocks, boulders, logs, and man–made objects
Associations with other life-forms: Known to have algae symbionts. In Canada *Ephydatia muelleri* colonies were most often associated with two sponges *Spongilla lacustris* and *Eunapius fragilis* and two bryozoans *Cristatella mucedo* Cuvier, 1798, and *Pectinatella magnifica* Leidy, 1851. Found growing on zebra mussels in the Great Lakes–St. Lawrence River system.
Ranges of abiotic variables:
    Alkalinity: Methyl orange: 30–130 ppm $CaCO_3$
    Calcium: 4.11–82 ppm
    Chlorine: 0.085–0.306
    CO_2: 3.5–12 ppm
    Conductivity: 30–921.6 μmhos/cm
    DO_2: 4.88–15.6 ppm
    Fluoride: 0.182–0.40 ppm
    HCO_3^- 12.2–15 ppm
    Magnesium: 0.4–70 ppm
    Nitrate: 0.431–0.486 ppm
    Oxidation Reduction Potential: −53.23–94.1 mV
    pH: 5.8–9.1
    Potassium: 0.31–8.58 ppm
    Salinity: 0.04–0.57 psu
    SiO_2: 0.04–11.6 ppm
    SO_4: 3.36–10.56 ppm
    Sodium: 0.39–0.79 ppm
    Temperature: 9–24° C
    Total Dissolved Solids: 0.05–0.74 0.431 ppm
    Total Hardness: 40–80 ppm $CaCO_3$

Ephydatia muelleri gemmules exposed to relatively low pH (5.8–6.5) resulted in reduced hatchability. Gemmules of *E. muelleri* may withstand long-term exposure to temperatures as low as −80° C.

In the United Kingdom *Ephydatia muelleri* was found frequently coexisting with *Eunapius fragilis* and associated with sites having lower than average salinity, conductivity, and total dissolved solids. *Ephydatia muelleri* is reported to have a patchy distribution across Europe.

Eunapius fragilis

Habitat: lotic and lentic waters
Substrate: rocks, logs, freshwater mussel shells
Associations with other life-forms: In Canada found to frequently occur with *Spongilla lacustris, Ephydatia muelleri* and bryozoans *Paludicella articulata* and *Pectinatella magnifica*. Reported coexisting with *Ephydatia muelleri* in UK. Parasitized by the spongillaflies *Sisyra apicalis, Sisyra vicaria*, and *Climacia areolaris* and sponge-eating caddisflies of the genus *Ceraclea*. In St. Lawrence River of eastern Canada and United States *Eunapius fragilis* is a common epizoic organism on unionid mussels.
Ranges of abiotic variables:
  pH: 4.2–9.
  Salinity: 0.04–0.57 psu
  Alkalinity: methyl orange: 14.7–170 ppm $CaCO_3$
  Alkalinity: phenolphthalein 0–6 ppm
  Ammonia Nitrogen: 0–0.5 ppm
  Calcium Hardness $CaCO_3$: 0–30 ppm
  Calcium: 1.6–45.6 ppm
  Chlorides: 29–239 ppm
  CO_2: 0–82 ppm
  Conductivity: 16–921.67 μmhos/cm
  DO_2: 2.5–17.97 ppm
  Magnesium Hardness $MgCO_3$: 4.3–120 ppm
  Magnesium: 1.05–11 ppm
  Nitrate Nitrogen: 0.001–0.385 ppm
  Nitrite Nitrogen: 0.001–0.232 ppm
  Oxidation Reduction Potential: −64.8–47 mV
  SiO_2: 0.3–24.9 ppm
  Sodium: 4.37–13.6 ppm
  Temperature: 11–34° C

In the United Kingdom *Eunapius fragilis* and *Ephydatia muelleri* were the most frequently occurring coexisting pair of species.

Heteromeyenia latitenta

Habitat: lotic waters

Substrate: rock

Associations with other life-forms: In the Pigeon River of Tennessee *Heteromeyenia latitenta* was always collected with *Radiospongilla cerebellata*.

Ranges of abiotic variables: From twenty-four samples taken at collection site on Pigeon River

DO_2: 3.6–9 ppm
Nitrates: 0.32–0.62 ppm
Nitrites: 0.002–0.095 ppm
pH: 6.3–8.4
Phosphates: 0.03–1.68 ppm
SiO_2: 2–11 ppm
Sulfates: 9–31 ppm
Total Ammonia: 0.8–3.01 ppm

Much remains to be learned concerning the ecology of this sponge.

Heteromeyenia tubisperma

Habitat: lotic waters

Substrate: rocks, wood

Associations with other life-forms: In Canada *Heteromeyenia tubisperma* was reported having associations with *Spongilla lacustris*, *Eunapius fragilis*, and *Ephydatia muelleri* and two bryozoans *Paludicella articulata* (Ehrenberg, 1831) and *Plumatella emarginata* Allman, 1844. Found growing on zebra mussels in the Great Lakes–St. Lawrence River system.

Ranges of abiotic variables:

Alkalinity (methyl orange): 110–170 ppm
Alkalinity (phenolphthalein): 0–2ppm
Calcium: 8–15 ppm
Free CO_2: 0.0–12.0 ppm
pH: 6.6–8.5
Temperature: 13.5–26.5 ppm
Total Hardness: 80.0–200.0 ppm

Heterorotula lucasi

Habitat: lotic waters
Substrate: rocks
Associations with other life-forms: Collected with *Radiospongilla cerebellata*.
Ranges of abiotic variables: From twenty-four samples taken at collection site on Pigeon River
- DO_2: 3.6–9 ppm
- Nitrates: 0.32–0.62 ppm
- Nitrites: 0.002–0.095 ppm
- pH: 6.3–8.4
- Phosphates: 0.03–1.68 ppm
- SiO_2: 2–11 ppm
- Sulfates: 9–31 ppm
- Total Ammonia: 0.8–3.01 ppm

Racekiela ryderi

Habitat: lotic waters; in Louisiana it is common in areas of acid drainages and is the most common species in habitats subject to seasonal drying.
Substrate: rocks, wood
Associations with other life-forms: Parasitized by the spongillaflies *Sisyra vicaria* and *Climacia areolaris*. Known to have algae symbionts.
Ranges of abiotic variables:
- Alkalinity methyl orangey: 4–128 ppm
- Alkalinity phenolphthalein: 0.06–0.27 ppm
- Ammonia nitrogen: 0.192
- Calcium Hardness (as $CaCO_3$): 24.5 ppm
- Calcium: 1.2–130 ppm
- Chloride: 1–5.3 ppm
- Chromium: 0.0004–0.057 ppm
- Conductivity: 37–170 μmhos/cm
- Copper: 0.09–0.45 ppm
- Dissolved oxygen: 7.1 ppm
- Free CO_2: 3.15–12 ppm
- Iron: < 0.002–0.033 ppm
- Magnesium: 1.05–150 ppm

Nitrate nitrogen: 0.001 ppm
pH: 4.2–8.5
Phosphate: 0.028–0.192 ppm
SiO_2: 3.05–13 ppm
Sulfate 1.5–5.06 ppm
Temperature: 19–32° C
Total Hardness (as $CaCO_3$): 22.5–80 ppm
Total solids: 164 ppm

Radiospongilla cerebellata

Habitat: lotic waters
Substrate: logs, rock
Associations with other life-forms: Known to have algae symbionts.
Ranges of abiotic variables: From 24 samples taken at collection site on Pigeon River
DO_2: 3.6–9 ppm
Nitrates: 0.32–0.62 ppm
Nitrites: 0.002–0.095 ppm
pH: 6.3–8.4
Phosphates: 0.03–1.68 ppm
SiO_2: 2–11 ppm
Sulfates: 9–31 ppm
Total Ammonia: 0.8–3.01 ppm

Radiospongilla crateriformis

Habitat: It has been reported by several researchers that the preferred habitat for *Radiospongilla crateriformis* is stagnant, turbid, alkaline waters. However, in Tennessee this species has been collected from rapid flowing streams and rivers.
Substrate: logs, wooden objects, rocks, boulders, and concrete structures. Can be found in temporary pools and ponds.
Associations with other life-forms: In Louisiana this sponge has been found to be parasitized by two spongillaflies: *Sisyra apicalis* and *Sisyra vicaria*.
Ranges of abiotic variables:
Calcium concentration: 130 PPM
DO_2: 4.3–8.8 ppm

Magnesium concentration: 150 ppm
Nitrates: 0.52–0.79 ppm
Nitrites: 0.001–0.60 ppm
pH: 7.0–8.2
Phosphates: 0.06–1.95 ppm
SiO_2: 2–11 ppm
Sulfates: 2–35 ppm
Total Ammonia: 0.89–2.06 ppm

Spongilla lacustris

Habitat: lotic and lentic waters, tolerant of brackish waters, semi-arid to permafrost areas, found at higher elevations than any other freshwater sponge in North America.

Substrata: rocks, logs, man-made objects

Associations with other life-forms: In Canada *Spongilla lacustris* was found commonly associated with *Eunapius fragilis* and *Ephydatia muelleri* and the bryozoan, *Paludicella articulata*. It commonly grows on shells of living unionid mussels. Known to have algae symbionts. Parasitized by the sponge eating caddisfly *Ceraclea resurgens*. A study in Wisconsin looked at algae and animal life found on *Spongilla lacustris*. Commensal algae, some of which were responsible for the green color of sponges, were *Cladophora*, *Spirogyra*, and diatoms of the genera *Gomphonema, Cocconeis, Navicula, Fragilaria, Cymbella, Rhopalodia,* and *Hantzschia*. Insect larvae found were caddisflies of the genera: *Hydropsyche, Cheumatopsyche, Chimarra, Limnephilus, Brachycentrus, Helicopsyche, Hydroptila, Molanna*; Stoneflies of the genera *Isoperla, Paragnetina, Acroneuria, Taeniopteryx, Hastaperla, Nemoura,* and *Pteronarcys*; Mayflies of the genera *Ephemerella, Heptagenia, Stenonema, Baetis,* and *Paraleptophlebia*, and the spongillafly *Climacia areolaris*. Found growing on zebra mussels in the Great Lakes–St. Lawrence River system.

Ranges of abiotic variables:
Alkalinity methyl orange: 90–170 ppm
Alkalinity phenolphthalein: 0–2 ppm
Ammonia nitrogen: < 0.001ppm
Calcium Hardness (as $CaCO_3$): 28.8 2–50 ppm
Calcium: 0.16–178 ppm

Chloride: 0–239 ppm
Conductivity: 9.4–687.67 μmhos/cm
Dissolved Solids: 50.06–670.76 ppm
DO_2: 2.4–15.6 ppm
Fluoride: 0.182–0.40 ppm
Free CO_2: 0–13.9 ppm
Iron: < 0.002–1.2 ppm
Magnesium Hardness: (as $MgCO_3$): 4.3 ppm
Magnesium: 0.40–64 ppm
Manganese: 0.2–64 ppm
Nitrate nitrogen: 0.192–9.9 ppm
Nitrite nitrogen: 0.001–9.95 ppm
Oxidation Reduction Potential: −19.05–47.07 mV.
pH: 4.2–9
Phosphate: 0.028–6.34 ppm
Potassium: 0.31–3.7 ppm
Salinity: 0.04–0.46 psu
SiO_2: 0–20.5 ppm
Sodium: 0.39–0.79 ppm
Sulfate: 5.06–37.0 ppm
Temperature 7.18–37° C
Total Hardness (as $CaCO_3$): 22.5–140 ppm
Total Dissolved solids: 20–164 ppm

In the United Kingdom salinity was a significant factor associated with the presence of *S. lacustris*. It preferred sites with lower salinity, conductivity, and total dissolved solids than *Eunapius fragilis* and *Ephydatia fluviatilis*. Also, in the UK its tolerance of lower water temperatures contributed to a more northern persistence, more frequently found above a latitude of 55° N than other UK species.

In Norway *S. lacustris* has been found in lakes located north of the Arctic Circle, up to 71° 03′ N. *Spongilla lacustris* preferred slightly acidic dystrophic lakes having rather low concentrations of calcium and magnesium.

Trochospongilla horrida

Habitat: lotic waters; reported by several researchers to occur in warm brackish water.

Substrate: rock
Associations with other life-forms: Parasitized by the spongillafly *Climacia areolaris*; known to have algae symbionts. Found growing on zebra mussels in the Great Lakes–St. Lawrence River system.
Ranges of abiotic variables:
    Alkalinity (methyl orange): 7–185 ppm
    Ammonia nitrogen: 0.005–0.047 ppm
    Calcium Hardness (as $CaCO_3$): 10.9–11 ppm
    Calcium: 4.1–60 ppm
    Chlorides: 4–< 15 ppm
    Conductivity: 0.068–322.0 μmhos/cm
    DO_2: 6.2–8 ppm
    Free Co_2: 0–15 ppm
    Magnesium Hardness (as $MgCO_3$): 5.4–7.1 ppm
    Magnesium: 1.31–50.0 ppm
    Nitrate nitrogen: 0.07–0.2 ppm
    Nitrite nitrogen: 0.001–< 0.007 ppm
    pH: 5.5–8.7
    Phosphate: 0.024–0.1 ppm
    SiO_2: 5–24.9 ppm
    Sulfate: 2.28–11.02 ppm
    Temperature: 18.5–34° C
    Total Hardness: 10.0–156.0 ppm
    Total Solids: 78.0–190.0 ppm
    Volatile Solids: 20–60 ppm

Trochospongilla leidyi

Habitat: lotic waters; reported by researchers to occur in warm brackish water
Substrate: rocks
Associations with other life-forms: Parasitized by the spongillaflies *Climacia areolaris* and *Climacia chapini* Parfin and Gurney, 1956.
Ranges of abiotic variables:
    Alkalinity Methyl orange: 14.7–185 ppm
    Alkalinity phenolphthalein: 15 ppm
    Ammonia nitrogen 0.009–0.03 ppm
    Calcium: 3–10 ppm

Conductivity: 80–3000 μmhos/cm
DO_2: 9–11 ppm
Free CO_2: 0–18
Magnesium: < 3–11 ppm
Nitrate nitrogen: 0.2–0.7 ppm
Nitrite nitrogen: 0.001–0.001
pH: 6.5–8.7
Phosphate: > 0.01–0.05
SiO_2: > 6.0–12 ppm
Sulfate: >1.0–10 ppm
Total Hardness: 10–< 50 ppm

Sources of water chemistry data: De Santo, E. M. and P. E. Fell, 1996; Evans, K. L, and D. J. S. Montagnes, 2019; Gugel, J. 2001; Harrison, F. W.,1974; Herrmann, S. J. et al., 2019; Jewell, M. F., 1935, and 1939; M. A. Poirrier 1970.

CHAPTER 8

Conservation

> "Of all the questions which can come before this nation,
> short of the actual preservation of its existence in a great war,
> there is none which compares in importance with the great
> central task of leaving this land even a better land
> for our descendants than it is for us."
>
> —THEODORE ROOSEVELT—

The biodiversity of freshwater ecosystems is remarkable. Freshwater organisms account for almost 6% of all described species, even though freshwater accounts for only 0.01% of the Earth's water and approximately 0.8% of the Earth's surface. Currently, no freshwater sponge is listed as threatened or endangered. To protect and maintain species diversity, anyone interested in the conservation of freshwater sponges should be involved with the conservation of freshwater environments. Organizations such as American Rivers, the Nature Conservancy, Trout Unlimited, American Whitewater, and others advocate clean and healthy waterways and develop policies and legislation toward that end. Through the actions of these groups and regulations of the Clean Water Act, streams once so polluted that water contact was unadvisable have been restored, allowing fishing, canoeing, and other aquatic recreational activities.

The Clean Water Act (CWA) of 1972 is the primary federal legislation regulating the discharge of pollutants into the waters of the United States. The development of the CWA stems from the passage of the Federal Water Pollution Control Act of 1948. Under the CWA, the Environmental Protection Agency (EPA) has regulatory control. The EPA has established water quality standards for industry and national standards for pollutants in surface waters. The CWA has been effective in reducing point-source pollution, such as effluent from a sewage treatment plant or factory. It has not been as successful when concerned with non-point pollution. Non-point pollution is more nebulous as pollution sources are from various places. Rain or melting snow results in runoff, containing various pollutants from the landscape.

Landowners having property boarding a creek or river can take direct action to improve water quality by maintaining, enhancing, or creating riparian habitat. Riparian habitats are the interface communities which transition from terrestrial to aquatic environments. Riparian vegetation is characterized by hydrophilic plants. Maintaining or restoration of riparian habitat is beneficial to the entire stream ecosystem. Adding to a wide range of functions and benefits, Riparian plant communities can:

1. Influence light quantity and quality, which affects production by algae symbionts and aquatic plants.
2. Act as a natural biofilter for capturing runoff.

Elk River, Franklin County, Tennessee.

3. Control erosion of streambanks and riverbanks, reducing sedimentation.
4. Provide shade and a cooling effect on water temperature. (The degree to which an absence of riparian vegetation can influence water temperature is significant. An increase to 100° F can occur removal.)
5. Provide substates for sponge attachment on exposed roots of trees and shrubs.
6. Provide organic materials such as leaves and fruits which serve as food for aquatic macroinvertebrates and release nutrients as they decompose.

Unfortunately, the loss of riparian vegetation is substantial. The US Fish and Wildlife Service estimates that 70% of the original riparian habitat within the United States has been altered or lost. Losses are due to several causes including agricultural practices, real estate development, construction of roads, the growth of towns and cities, and others.

Funding from conservation programs that maintain and create riparian habitat is making a difference. The Nature Conservancy offers financial incentives to eligible landowners through the Conservation Reserve Enhancement Program (CREP) for the purpose of creating riparian zones. The Tennessee Department of Agriculture and the University of Tennessee Institute of Agriculture have a Community Riparian Restoration Program that strives to connect people and communities with their watersheds by raising awareness and understanding of riparian areas. Additionally, the US Fish and Wildlife Service collaborates with citizen groups and other nonprofit organizations to implement riparian restoration projects.

Equally important is watershed vegetation. A watershed is an area of land and the water it captures and transfers to creeks and rivers. Watershed vegetation is important for retention of soil moisture and reducing sediment loading of streams. This vegetation slows the movement of water across the landscape, thereby reducing soil erosion and sediment accumulation.

Some sedimentation is normal and important to the operation of streams. Excessive sedimentation can be detrimental to freshwater sponges. Incurrent canals can be occluded, thereby preventing or restricting water flow through a sponge. Sponges may be buried or smothered by sediments. Reproductive success can be reduced as sediments prevent the attachment of larvae and gemmules to substrates. Photosynthesis by symbiotic algae is reduced.

Human modifications and climate change influence streamflow. Streams and rivers have undergone anthropogenic modifications for flood control, irrigation, generation of electricity, navigation, and water sources for municipalities. Climate influences streamflow by determining air temperature, when precipitation events occur, whether precipitation is rain or snow, and timing of snowmelt. Climate change is affecting precipitation events through changes in the water cycle. Historical patterns of rainfall are being disrupted by a lack of or a drastic increase in precipitation. This disruption is causing drought and flood events to occur with greater frequency.

The United States Geological Survey (USGS) monitors streamflow at thousands of sites across the nation. Streamflow data gathered from 1980 to 2014 have shown that lows and highs of streamflow have changed from their normal patterns. Obviously, streamflow varies over the course of a year, historically higher in the spring and lower in the fall and is determined by precipitation events. Data collected by the USGS reveal both low and high streamflow events occur more frequently but for a shorter duration than normal. High flow rates also occur at lower magnitudes. Changes from the normal streamflow patterns can be problematic for aquatic species because over time species become fine-tuned to environmental regimes, such as when the rainy and dry seasons occur. Reproduction may occur at the normal time but changes in water flow may cause a reduction in fecundity, specifically survivability of gametes, eggs, larvae, and young, thereby reducing a population's ability to survive.

Aquatic organisms face difficult conditions during droughts. Low water levels, reduced stream flow and complete dry-up of small creeks and vernal pools will lead to desiccation and death. Sponge dispersal is hindered by the loss of connections between streams. Low water flow will lead to higher concentration of pollutants. As water temperature rises, dissolved oxygen concentration decreases which can lead to a blue-green algae bloom and a release of toxins. Loss of habitat occurs as substrate becomes exposed to the atmosphere as water levels decline.

On the other hand, extreme rainfall events will result in excessive runoff. In agricultural and industrial areas runoff can transfer herbicides, pesticides, fertilizer, and other pollutants to creeks and rivers. Stream bank erosion will increase resulting in greater sedimentation. Sewage treatment facilities could be overwhelmed allowing raw sewage to enter waterways.

Temperatures of lakes, streams, and rivers are increasing. A National Aeronautics and Space Administration (NASA) analysis of more than 25 years of satellite temperature data and ground measurements of 235 lakes on six continents found lakes are warming an average of 0.34° C each decade, which is greater than the warming rate of either the ocean or the atmosphere. Many aquatic organisms have adapted to live within specific temperature ranges.

Increasing water temperatures affect aquatic organisms by lowering survival rates, reduction in reproductive success, and death. Lethal temperatures have been determined for several groups of aquatic insect groups. Differences exist among taxonomic orders and among the species comprising an order. Based on combining the results of several studies the order Ephemeroptera, mayflies, have an average lethal temperature of 22.3° C (71.8° F) while for Trichoptera, caddisflies, the average lethal temperature was 30.1° C (86.2° F). Of course, there was significant within-order variation by species. Thermal stress can limit reproductive success by decreasing sperm and egg production through the reabsorption of spermatic cysts and oocytes and cause the release of asexual propagules when environments are unfavorable for their survival. Elevated temperatures can result in the increase of heat shock proteins, which results in rapid activation of genes involved in repairing cellular damage. Temperature influences sponge feeding behavior by increasing or decreasing water filtration rates and by decreasing choanocyte chamber density and size. Changes will occur in the species composition of aquatic biological communities.

Some physical properties of water, such as density and dissolved oxygen concentration, are influenced by temperature. Photosynthesis and diffusion from the atmosphere are the two sources of oxygen for bodies of water. As water warms, its density is lowered. Dissolved oxygen (DO) concentrations have an inverse relationship with temperature, as temperature rises DO declines. Lower DO concentrations result in a reduction of growth for many aquatic organisms while at the same time promoting toxic blooms by cyanobacteria (a.k.a., blue-green algae).

Some of the consequences of increasing water temperature are:

1. Cool water species will encounter invasion of their home waters by warm water species.
2. Cool water species could face population decline or extinction, not only due to their environments warming, but also from competition with warm water species.
3. Changes in food web composition and energy transfer will occur as species become extinct or populations decline.
4. Will see an increase in the frequency and duration of droughts and rain events.
5. The ability of some ecosystems to modify the impacts of storm surges and floodwater will be altered.
6. The concentration of pollutants will increase as water levels of streams and lakes decline.

7. The spread of pathogens, parasites, and disease.
8. Changes in microbial composition of sponge microbiomes.

Rising sea levels, droughts, and greater human demands placed on water resources will increase the salinity of coastal freshwater rivers, streams, and marshes, especially in low-lying areas. Increases in salinity of freshwater bodies will harm aquatic plants and animals, forcing those that can to move inland and those that can't to adapt or die.

Disease may be another consequence of climate change. Elevated temperatures and agricultural runoff containing herbicides and pesticides can reduce the fitness of aquatic organisms. When sponges encounter adverse environmental conditions, normal sponge–bacterial associations can become destabilized, possibly resulting in the decline of a sponge's fitness. The decline in fitness is a result of major shifts in the taxonomic composition of a sponge's microbiome. A microbiome is the collection of all microbes such as bacteria, viruses, fungi, and their genes that naturally live on an organism. Changes in the microbiomes of other animals, including humans, are known to result in a loss of fitness. Evidence shows that the microbiome can influence key life processes of the host, such as protection from pathogens. Conceivably the biggest threat of climate change to freshwater sponges will be the changes that occur within the microbial community comprising their microbiomes.

Elevated water temperatures associated with climate change will have a considerable influence on microbial composition of microbiomes and sponge fitness. Some sea sponges have been affected by elevated temperatures. Even slight temperature increases can be disruptive. In the sea sponge *Halichondria bowerbanki* Burton, 1930, an increase of just 1–2º C above ambient resulted in changes in the microbial composition of the microbiome, as some microbes disappeared and were replaced by new microbes.

Lake Baikal, located in southern Siberia, is the largest freshwater lake on planet Earth, containing about 20% of the world's surface fresh water. In Lake Baikal, diseased individuals of the sponge *Lubomirskia baikalensis* were found to be dominated by *Synechococcus*, a unicellular cyanobacterium, and Verrucomicrobia, a little studied phylum of gram-negative bacteria, when compared to healthy individuals. The researchers believed increased methane production within the lake caused shifts in the microbiomes of diseased sponges. Another study involving cell cultures of *L. baikalensis* found mass mortality of symbiotic Chlorophyta (green algae) associated with shifts of the sponge's microbiome.

A study of the microbiome of *Ephydatia muelleri* in the Sooke, Nanaimo, and Cowichan rivers on Vancouver Island, British Columbia found it to be dominated by *Sediminibacterium*, a purple sulfur bacterium, *Comamonas*, and an undescribed bacterium of the order Rhodospirillales. Interestingly, the microbial composition of the microbiome was different from that of surrounding water and nearby biofilms. Also, the microbiomes of *E. muelleri* from the three rivers were different in taxonomic composition. The researchers reported the microbiome of *E. muelleri* to share many genomic similarities with microbiomes of saltwater sponges. Similarities included having an abundance of defense-related proteins and genes for production of vitamin B12.

Several studies of sea sponge diseases associated with elevated water temperatures have been conducted. However, the same cannot be said concerning diseases of freshwater sponges. The few studies dealing with microbiomes of freshwater sponges have found them to be highly diverse. This is an aspect of freshwater sponge biology in need of study. Investigations of possible diseases of freshwater sponges resulting from shifts in their microbiomes, due to climate change, should be undertaken.

A recent discovery in China gives hope to the current situation of global temperature changes. This discovery suggests sponges not only survived the Ordovician mass extinction, caused by the abrupt onset of an ice age, but after which thrived. Scientists have found a large diverse assemblage of sponge fossils in Anji Biota, a fossil deposit in Zhejiang Province. This finding suggests sponges may have dominated the seas after the Ordovician extinction until significant recovery of other benthic organisms. Researchers believe sponges thrived because of their ability to tolerate a drastic decline in water temperature and anoxia. Most other benthic species lacked this ability, and their populations became extinct. The deaths of these organisms resulted in an enormous increase of nutrients for sponges in the form of particulate organic matter.

Environmental extremes experienced in the temperate climate zone stress freshwater sponges, especially water physiochemical conditions. Aquatic extremes include spring floods, summer drought, and winter ice. How will freshwater sponges adapt? Duplicated genes will play a significant role as freshwater sponges adapt to changing environments.

Gene duplication is an important mechanism by which organisms acquire new genes. Despite being anatomically simple, sponges have nearly twice as many genes as other animals. Gene duplication is responsible for this large amount of genetic material. Duplicate genes provide new genetic material for mutations, genetic drift, and natural selection to act upon and are free to evolve independently from their copy. This can result in genetic diversity. This strategy will prove critical to freshwater sponges as they are forced to adapt to climate change stressors.

Duplicated genes are most frequently nonfunctional but can become functional. In most cases, the duplicated genes acquire degenerative mutations and become known as pseudogenes, which are nonfunctional. However, if the duplicated gene undergoes neofunctionalization and acquires a new function that was not present in the gene from which it was copied the gene becomes a new gene. New genes can provide an organism with genetic plasticity. Genetic plasticity is a specific gene or genotype expressing different phenotypes in different environments. It provides an organism with the ability to adapt to changing environments by providing it with new abilities. Without gene duplication, the ability to adapt to changing environments would be very difficult.

The effects of climate change on aquatic organisms are already evident and will accelerate if we do nothing. Our work to reverse climate change and to have clean, healthy waters is necessary if the diversity of aquatic life is to be maintained.

Climate change and its associated issues have created major challenges for freshwater sponges and other aquatic organisms. However, there are other serious threats to freshwater biodiversity. One that is quite serious is the presence of minute particles of plastic in terrestrial and aquatic environments. Micro and nano plastic (MNPs) pollution is of major concern to biologists. Microplastics and nanoplastics are ubiquitous in terrestrial and aquatic ecosystems. These plastics are prevalent in terrestrial, marine, and freshwater environments and are an endangerment to aquatic organisms. Studies have proven they pose significant threats to microscopic biota such as green algae species of the genus *Chlorella*, as well as copepods, fish, and other organisms. *Chlorella* species are known to be endosymbionts of freshwater sponges. *Chlorella* has a high photosynthesis efficiency, about 8%. Photosynthetic efficiency is the fraction of light energy converted into chemical energy during photosynthesis. Nanoplastics are known to lower the photosynthetic efficiency of *Chlorella* species, resulting in lower glucose production. Lower photosynthetic efficiency equates to less dissolved oxygen. As dissolved oxygen concentration drops in freshwaters, the likelihood of a toxic cyanobacteria bloom increases. Blooms affect food chains through the release of cyanotoxins. These toxins can disrupt the development and fertility of aquatic organisms, as well as causing their death.

Periphyton accumulates and retains nanoplastics. Freshwater sponges consume periphyton when it becomes dislodged from its substrate. Studies are needed to investigate the effects of nanoplastics on the physiology of freshwater sponges.

APPENDIX A

Suggestions for Teachers and Researchers

This book provides aspects of the biology of freshwater sponges. Topics concerning anatomy, ecology, life history, taxonomy, and identification have been covered. Hopefully, we have generated an awareness and an interest in freshwater sponges.

Teachers and students can use this book as a starting point for developing educational and research projects to further advance our knowledge of freshwater sponges.

Despite the fact freshwater sponges are important benthic invertebrates performing important roles in maintaining the structure and functioning of freshwater ecosystems, data necessary to conserve and manage freshwater sponges have not been collected. Even the most basic information, such as which species are present, where are they located, and how many individuals are present, is unavailable to most, if not all, state and federal agencies responsible for conservation of freshwater organisms.

We encourage educators and researchers interested in freshwater sponges to consider developing teaching and research projects that focus on sponges. We offer the following as examples of projects to get you started in working with freshwater sponges.

1. Teachers can utilize freshwater sponges as models for teaching various topics of biology. Doing so will quickly raise awareness of freshwater sponges.
2. Conduct stream or river surveys documenting sponge species present, their distributions, and numbers of individuals. The range maps presented in this book reveal the lack of knowledge concerning the distributions of sponges in Tennessee. This situation is true for other states.
3. Survey lakes, ponds, and Tennessee Valley Authority (TVA) reservoirs for sponges. Only two, very limited in scope, surveys of lentic waters have

occurred in Tennessee. The first at Reelfoot Lake in 1942 and the second in 2014 in a single cove of Chilhowee Lake existing within the boundary of the Great Smoky Mountains National Park.
4. Consider taking on long-term studies to obtain temporal trend data concerning population dynamics and changes in sponge community structure. This is an area of great need as information is very limited.
5. Consider long-term studies of a specific stream to document the effects of natural and anthropogenic events on sponge populations and communities.
6. Determine the microscopic life-forms of the microbiomes of freshwater sponges and the influence increasing water temperatures might have on these organisms.
7. Are there freshwater sponge diseases associated will increasing temperatures and changes in microbiome?
8. View the intestines of crayfish and minnows and darters for the presence of spicules and gemmules (this one could easily be conducted by high school students or college undergraduates).
9. Investigate the prevalence of plastic particulates in aquatic food chain, especially at the producer level, and their prevalence in sponges and other aquatic organisms. Are nanoplastics disruptive to sponge growth and reproduction?

These projects are just a few of the many that could be conducted. These are offered as examples of projects that could be used to introduce students and others to freshwater sponges.

APPENDIX B

Distributions of Freshwater Sponge Species in the United States

TABLE B1. Distribution of Freshwater Sponge Species by State

Sponge	State
Anheteromeyenia argyrosperma	Arkansas Connecticut Delaware Florida Illinois Indiana Louisiana Massachusetts Michigan Mississippi New Jersey Pennsylvania South Carolina Virginia Wisconsin
Cherokeesia armata	Tennessee
Corvomeyenia carolinensis	Connecticut South Carolina
Corvomeyenia everetti	Delaware Massachusetts Michigan Wisconsin
Corvospongilla becki	Alabama Louisiana Tennessee
Corvospongilla novaeterrae	Connecticut Michigan

Sponge	State
Dosilia palmeri	Alabama California Florida Louisiana New Mexico Texas
Dosilia radiospiculata	Alabama Arizona Arkansas California Illinois Louisiana Mississippi Ohio Oklahoma Texas Virginia
Duosclera mackayi	Connecticut* Florida Georgia Louisiana Massachusetts Michigan New Jersey New York Wisconsin
Ephydatia fluviatilis	Colorado Florida Illinois Indiana Kentucky* Louisiana Massachusetts Michigan Montana North Dakota Ohio Pennsylvania South Dakota Tennessee Utah Wisconsin
Ephydatia millsi	Florida

Sponge	State
Ephydatia muelleri	Arizona Arkansas Colorado Connecticut District of Columbia Illinois Indiana Iowa Maryland Massachusetts Michigan Minnesota Montana Nebraska(?) New York Pennsylvania Tennessee Virginia Wisconsin Wyoming
Ephydatia subtilis	Florida
Eunapius fragilis	Alabama Arkansas Colorado Connecticut Delaware Florida Illinois Indiana Iowa Kansas Kentucky Louisiana Maine Massachusetts Michigan Minnesota Mississippi Montana New Jersey New York North Dakota Oklahoma Pennsylvania

Sponge	State
Eunapius fragilis	South Carolina South Dakota Tennessee Texas Washington Wisconsin Wyoming
Heteromeyenia baileyi	Delaware Hawaii Illinois Indiana Louisiana Massachusetts Montana New Jersey New York Pennsylvania South Carolina South Dakota Texas Wisconsin
Heteromeyenia latitenta	Arkansas Illinois Indiana Kentucky* Massachusetts New Jersey New York Ohio Pennsylvania Tennessee
Heteromeyenia longistylis	Pennsylvania
Heteromeyenia tentasperma	New Jersey New York Ohio Pennsylvania Wisconsin
Heteromeyenia tubisperma	California Florida Illinois Indiana Iowa Kansas

Sponge	State
Heteromeyenia tubisperma	Michigan New Jersey New York Ohio South Dakota Tennessee Wisconsin
Heterorotula lucasi	Tennessee
Pottsiela aspinosa	Florida Indiana Massachusetts Michigan New Jersey Oklahoma Virginia
*Racekiela biceps***	Michigan
Racekiela ryderi	Connecticut Delaware Florida Georgia Indiana Iowa Kentucky* Louisiana Massachusetts Michigan Mississippi New Hampshire New Jersey New York North Carolina Pennsylvania South Carolina Tennessee Texas Virginia Wisconsin
*Radiospongilla cerebellata***	Alabama Louisiana Tennessee Texas

Sponge	State
Radiospongilla crateriformis	Alabama Delaware District of Columbia Florida Illinois Indiana Kentucky Louisiana Maryland Massachusetts Mississippi New York Ohio Oklahoma Pennsylvania South Carolina Tennessee Texas Wisconsin
*Spongilla alba***	Florida Louisiana
Spongilla cenota	Florida Texas
Spongilla lacustris	Alaska Colorado Connecticut Delaware District of Columbia Florida Georgia Idaho Illinois Indiana Iowa Kansas Kentucky Louisiana Maine Maryland Massachusetts Michigan Minnesota Mississippi

Sponge	State
Spongilla lacustris	Missouri Montana New Jersey New York Ohio Oklahoma Pennsylvania South Carolina Tennessee Texas Washington Wisconsin Wyoming
Spongilla wagneri	Florida Louisiana South Carolina
Stratospongilla penneyi	Florida
Trochospongilla horrida	Arkansas Delaware Florida Illinois Kentucky Maine Massachusetts Mississippi New Jersey Pennsylvania South Carolina Tennessee Texas
Trochospongilla leidyi	Arkansas Florida Illinois Kentucky Louisiana New Jersey Ohio Oklahoma Pennsylvania Tennessee Texas

Sponge	State
Trochospongilla pennsylvanica	Connecticut Delaware Florida Massachusetts Michigan Mississippi New Jersey Ohio Pennsylvania South Carolina Virginia Wisconsin

*Identified from spicules found in sediments
** Species under scrutiny as to identification
(?) Identified to the genus *Ephydatia*, researchers though it matched best with *Ephydatia muelleri*
The above list should not be considered all inclusive.

TABLE B2. State Lists of Freshwater Sponges

State	Sponge
Alabama	*Corvospongilla becki* *Dosilia palmeri* *Dosilia radiospiculata* *Eunapius fragilis* *Radiospongilla cerebellata*** *Radiospongilla crateriformis*
Alaska	*Spongilla lacustris*
Arizona	*Dosilia radiospiculata* *Ephydatia muelleri*
Arkansas	*Anheteromeyenia argyrosperma* *Dosilia radiospiculata* *Ephydatia muelleri* *Eunapius fragilis* *Heteromeyenia latitenta* *Trochospongilla horrida* *Trochospongilla leidyi*
California	*Dosilia palmeri* *Dosilia radiospiculata* *Heteromeyenia tubisperma*
Colorado	*Ephydatia fluviatilis* *Ephydatia muelleri* *Eunapius fragilis* *Spongilla lacustris*

State	Sponge
Connecticut	*Anheteromeyenia argyrosperma* *Corvomeyenia carolinensis* *Corvospongilla novaeterrae* *Duosclera mackayi** *Ephydatia muelleri* *Eunapius fragilis* *Racekiela ryderi* *Spongilla lacustris* *Trochospongilla pennsylvanica*
Delaware	*Anheteromeyenia argyrosperma* *Corvomeyenia everetti* *Eunapius fragilis* *Heteromeyenia baileyi* *Racekiela ryderi* *Radiospongilla crateriformis* *Spongilla lacustris* *Trochospongilla horrida* *Trochospongilla pennsylvanica*
District of Columbia	*Ephydatia muelleri* *Radiospongilla crateriformis* *Spongilla lacustris*
Florida	*Anheteromeyenia argyrosperma* *Dosilia palmeri* *Duosclera mackayi* *Ephydatia fluviatilis* *Ephydatia millsii* *Ephydatia subtilis* *Eunapius fragilis* *Heteromeyenia tubisperma* *Pottsiela aspinosa* *Racekiela ryderi* *Radiospongilla crateriformis* *Spongilla alba*** *Spongilla cenota* *Spongilla lacustris* *Spongilla wagneri* *Stratospongilla penneyi* *Trochospongilla horrida* *Trochospongilla leidyi* *Trochospongilla pennsylvanica*
Georgia	*Duosclera mackayi* *Racekiela ryderi* *Spongilla lacustris*

State	Sponge
Hawaii	*Heteromeyenia baileyi*
Idaho	*Spongilla lacustris*
Illinois	*Anheteromeyenia argyrosperma* *Dosilia radiospiculata* *Ephydatia fluviatilis* *Ephydatia muelleri* *Eunapius fragilis* *Heteromeyenia baileyi* *Heteromeyenia latitenta* *Heteromeyenia tubisperma* *Radiospongilla crateriformis* *Spongilla lacustris* *Trochospongilla horrida* *Trochospongilla leidyi*
Indiana	*Anheteromeyenia argyrosperma* *Ephydatia fluviatilis* *Ephydatia muelleri* *Eunapius fragilis* *Heteromeyenia baileyi* *Heteromeyenia latitenta* *Heteromeyenia tubisperma* *Pottsiela aspinosa* *Racekiela ryderi* *Radiospongilla crateriformis* *Spongilla lacustris*
Iowa	*Ephydatia muelleri* *Eunapius fragilis* *Heteromeyenia tubisperma* *Racekiela ryderi* *Spongilla lacustris*
Kansas	*Eunapius fragilis* *Heteromeyenia tubisperma* *Spongilla lacustris*
Kentucky	*Ephydatia fluviatilis** *Eunapius fragilis* *Heteromeyenia latitenta** *Racekiela ryderi* *Radiospongilla crateriformis* *Spongilla lacustris* *Trochospongilla horrida* *Trochospongilla leidyi*

State	Sponge
Louisiana	*Anheteromeyenia argyrosperma* *Corvospongilla becki* *Dosilia palmeri* *Dosilia radiospiculata* *Duosclera mackayi* *Ephydatia fluviatilis* *Eunapius fragilis* *Heteromeyenia baileyi* *Racekiela ryderi* *Radiospongilla cerebellata*** *Radiospongilla crateriformis* *Spongilla alba*** *Spongilla lacustris* *Spongilla wagneri* *Trochospongilla horrida* *Trochospongilla leidyi*
Maine	*Anheteromeyenia argyrosperma* *Eunapius fragilis* *Spongilla lacustris* *Trochospongilla horrida*
Maryland	*Ephydatia muelleri* *Radiospongilla crateriformis* *Spongilla lacustris*
Massachusetts	*Anheteromeyenia argyrosperma* *Corvomeyenia everetti* *Duosclera mackayi* *Ephydatia fluviatilis* *Ephydatia muelleri* *Eunapius fragilis* *Heteromeyenia baileyi* *Heteromeyenia latitenta* *Heteromeyenia tubisperma* *Pottsiela aspinosa* *Racekiela ryderi* *Radiospongilla crateriformis* *Spongilla lacustris* *Trochospongilla horrida* *Trochospongilla pennsylvanica*
Michigan	*Anheteromeyenia argyrosperma* *Corvomeyenia everetti* *Corvomeyenia novaeterrae* *Duosclera mackayi* *Ephydatia fluviatilis* *Ephydatia muelleri* *Eunapius fragilis*

State	Sponge
Michigan	*Heteromeyenia tubisperma* *Pottsiela aspinosa* *Racekiela ryderi* *Spongilla lacustris* *Trochospongilla pennsylvanica*
Minnesota	*Anheteromeyenia argyrosperma* *Ephydatia muelleri* *Eunapius fragilis* *Spongilla lacustris* *Trochospongilla pennsylvanica*
Mississippi	*Anheteromeyenia argyrosperma* *Dosilia radiospiculata* *Eunapius fragilis* *Racekiela ryderi* *Radiospongilla crateriformis* *Spongilla lacustris* *Trochospongilla horrida* *Trochospongilla pennsylvanica*
Missouri	*Spongilla lacustris*
Montana	*Ephydatia fluviatilis* *Ephydatia muelleri* *Eunapius fragilis* *Heteromeyenia baileyi* *Spongilla lacustris*
Nebraska	*Ephydatia muelleri*(?)
Nevada	No records found
New Hampshire	*Racekiela ryderi*
New Jersey	*Anheteromeyenia argyrosperma* *Duosclera mackayi* *Eunapius fragilis* *Heteromeyenia baileyi* *Heteromeyenia latitenta* *Heteromeyenia tentasperma* *Heteromeyenia tubisperma* *Pottsiela aspinosa* *Racekiela ryderi* *Spongilla lacustris* *Trochospongilla horrida* *Trochospongilla leidyi* *Trochospongilla pennsylvanica*
New Mexico	*Dosilia palmeri*

State	Sponge
New York	*Duosclera mackayi* *Ephydatia muelleri* *Eunapius fragilis* *Heteromeyenia baileyi* *Heteromeyenia latitenta* *Heteromeyenia tentasperma* *Heteromeyenia tubisperma* *Racekiela ryderi* *Radiospongilla crateriformis* *Spongilla lacustris*
North Carolina	*Racekiela ryderi*
North Dakota	*Ephydatia fluviatilis* *Eunapius fragilis*
Ohio	*Dosilia radiospiculata* *Ephydatia fluviatilis* *Heteromeyenia latitenta* *Heteromeyenia tentasperma* *Heteromeyenia tubisperma* *Radiospongilla crateriformis* *Spongilla lacustris* *Trochospongilla leidyi* *Trochospongilla pennsylvanica*
Oklahoma	*Dosilia radiospiculata* *Eunapius fragilis* *Pottsiela aspinosa* *Radiospongilla crateriformis* *Spongilla lacustris* *Trochospongilla leidyi*
Oregon	No records found
Pennsylvania	*Anheteromeyenia argyrosperma* *Ephydatia fluviatilis* *Ephydatia muelleri* *Eunapius fragilis* *Heteromeyenia baileyi* *Heteromeyenia latitenta* *Heteromeyenia longistylis* *Heteromeyenia tentasperma* *Racekiela ryderi* *Radiospongilla crateriformis* *Spongilla lacustris* *Trochospongilla horrida* *Trochospongilla leidyi* *Trochospongilla pennsylvanica*

State	Sponge
Rhode Island	No records found
South Carolina	*Anheteromeyenia argyrosperma* *Corvomeyenia carolinensis* *Eunapius fragilis* *Heteromeyenia baileyi* *Racekiela ryderi* *Radiospongilla crateriformis* *Spongilla lacustris* *Spongilla wagneri* *Trochospongilla horrida* *Trochospongilla pennsylvanica*
South Dakota	*Ephydatia fluviatilis* *Eunapius fragilis* *Heteromeyenia baileyi* *Heteromeyenia tubisperma*
Tennessee	*Cherokeesia armata* *Corvospongilla becki* *Ephydatia fluviatilis* *Ephydatia muelleri* *Eunapius fragilis* *Heteromeyenia latitenta* *Heteromeyenia tubisperma* *Heterorotula lucasi* *Racekiela ryderi* *Radiospongilla cerebellata*** *Radiospongilla crateriformis* *Spongilla lacustris* *Trochospongilla horrida* *Trochospongilla leidyi*
Texas	*Dosilia palmeri* *Dosilia radiospiculata* *Eunapius fragilis* *Heteromeyenia baileyi* *Racekiela ryderi* *Radiospongilla cerebellata*** *Radiospongilla crateriformis* *Spongilla cenota* *Spongilla lacustris* *Trochospongilla horrida* *Trochospongilla leidyi*
Utah	*Ephydatia fluviatilis*
Vermont	No records found

State	Sponge
Virginia	*Anheteromeyenia argyrosperma* *Dosilia radiospiculata* *Ephydatia muelleri* *Pottsiela aspinosa* *Racekiela ryderi* *Trochospongilla pennsylvanica*
Washington	*Eunapius fragilis* *Spongilla lacustris*
West Virginia	No records found
Wisconsin	*Anheteromeyenia argyrosperma* *Corvomeyenia everetti* *Duosclera mackayi* *Ephydatia fluviatilis* *Ephydatia muelleri* *Eunapius fragilis* *Heteromeyenia baileyi* *Heteromeyenia tentasperma* *Heteromeyenia tubisperma* *Racekiela ryderi* *Radiospongilla crateriformis* *Spongilla lacustris* *Trochospongilla pennsylvanica*
Wyoming	*Ephydatia muelleri* *Eunapius fragilis* *Spongilla lacustris*

* Identified from spicules found in sediments.
** Species under scrutiny as to identification.
(?) Identified to the genus Ephydatia, researchers thought it matched best with *Ephydatia muelleri*.
The above list should not be considered all inclusive.

GLOSSARY

Acantha: to have spines.
Acanthoxea: a spicule that's spiny rod–shaped and has pointed ends.
Alveolate: to have pits or be pitted.
Amniotic: a reference to the amnion which is a thin membrane enclosing the embryos or fetuses of reptiles, birds, and mammals.
Amphioxea: a spicule that's slightly curved needlelike, pointed at both ends.
Amoebocyte: a cell having the characteristics of an amoeba, or the cell is capable of amoeba–like movement.
Amphistrongyla: a slightly curved sponge spicule rounded at both ends.
Aniso: to be unequal.
Anisotropic: to have different values when measured in different directions.
Anoxia: the absence of oxygen.
Aperture: an opening or hole.
Apopyle: pore or opening though which water leaving a choanocyte chamber passes.
Archaeocyte: a totipotent amoeboid–like cell having several functions within a sponge and transform to become other cell types.
Asconoid: a sponge having no invaginations or chambers in its water canal system.
Asexual reproduction: reproduction of a single parent resulting in the production of progeny that are genetic clones of the parent.
Asymmetrical: lacking symmetry.
Autotroph: an organism capable of using solar energy in the process of photosynthesis or chemical energy in chemosynthesis to form energy–rich organic molecules through the incorporation of carbon into simple inorganic compounds.
Basal Spongin Plate: a plate-like structure by which a sponge attaches to a substrate, may also have gemmules attached.
Bauplan: structural body design.
Benthic: an ecological region, the substrate of a body of water, or a reference to organisms living there.
Benthos: flora or fauna found living on or near the bottom of a body of water
Binominal nomenclature: using two–words, genus and specific epithet, to name an organism.
Biodiversity: the variety of life-forms living in a specified area.
Biofilm: a thick layer of prokaryotic organisms that aggregate to form a colony which attaches to a substrate by a polysaccharide (slime) layer.
Biogeographical realm: a large biogeographic division of the Earth's land surface, based on distributional patterns of organisms.
Biological Clock: the innate mechanism that controls rhythmic physiological activities of an organism.

Biomass: the total weight of organism per unit area, may be measured as weight (grams of sponge per square meter) or mass (bushels of corn per acre).

Birotule: a spicule having a shaft with disc-like ends.

Bivalve: to have two shells (valves), refers to mussels, clams, oysters.

Bloom: an explosive, very rapid growth of a population.

Botryoidal: in the form of a cluster of grapes.

Cambrian Period: a 56-million-year interval from the end of the preceding Ediacaran Period 541 million years ago to the beginning of the Ordovician Period 485 million years ago. An important time period for animal evolution. Based on fossil records this is the time period during which most of the major groups (Phyla) of animals first appeared on Earth. An event known as the "Cambrian Explosion."

Carboniferous Period: a 60-million-year interval from the end of the Devonian Period 359 million years ago, to the beginning of the Permian Period, 299 million years ago. Named carboniferous because coal deposits represent the remains of forests from this period.

Chironomids: midges of the family Chironomidae, nonbiting midges comprising a family within the suborder Nematocera.

Chitin: A derivative of glucose. Used as a filament on which sclerocytes secrete silica dioxide to form a spicule.

Choanocytes: the collar cells that form the choanoderm, contain a central flagellum surrounded by a collar of microvilli, capture food and maintain water flow through the sponge body.

Choanocyte chamber: a deep pouch lined with choanocytes, found in leuconoid sponges.

Choanoderm: a layer of cells composed of choanocytes, where food is captured, and water flow is facilitated.

Choanosomal skeleton: a skeleton supporting the canal system consisting of a dense, irregular mesh of polygons formed by secondary fibers and primary fibers rise from it.

Cirrus: a tendril-like structure found on the foramen of sponges of the genus *Heteromeyenia*.

Clone: an individual which is a genetic replica of its single parent, resulting from asexual reproduction.

Coccus: to be circular or round.

Collagen: a structural protein in sponges which exists as a jelly–like substance or fibers, forms fibrous network of the sponge skeleton.

Commensal: a reference to commensalism, a relationship between two organisms in which one benefits and the other is unaffected.

Cross-fertilization: the union of gametes from two individuals

Cryptobiosis: a physiological state in which metabolic activity is reduced to an almost undetectable level. All measurable metabolic processes stop, preventing reproduction and development.

Ctenophora: a phylum of marine invertebrates which use bands of cilia for swimming, referred to as "combs" or "ctenes."

Dendritic: a tree–like structure, to have branches.

GLOSSARY

Desiccation: the loss of water, process of drying out, dehydration.

Diagenetic: from diagenesis, is the process by which sediments are altered after they are deposited, but before becoming rock, or a recombination or arrangement of components of molecules or compounds resulting in a new product.

Diapause: a state of dormancy, suspended or delayed development controlled by endogenous factors.

Dichotomous Key: a decision tool which provides two choices, one is true the other false concerning an unidentified organism.

Diploid: to have two sets of chromosomes (2n).

Diptera: an order of insects that comprises the two-winged or true flies, their hind wings are reduced to form balancing organs.

Ecosystem: a biological community formed by the interaction of organisms with their physical environment.

Ectosomal skeleton: skeleton of spicules and fibers found at the surface (epidermis), in that section of the sponge in which there is no choanoderm.

Endogenous: originating from within an organism.

Endosymbiotic: referring to an organism living within the body or cell of another organism.

Eocene: an epoch lasting from 56 to 34 million years ago, spans the time from the end of the Paleocene Epoch to the beginning of the Oligocene Epoch.

Ephemeral: a temporary or short period of time.

Epifauna: an animal living on the surface of another animal or plant.

Estivation: a time of prolonged torpor or dormancy of an animal during hot, dry, or cold periods.

Eutrophic: to have adequate to high levels of nutrients, a highly productive aquatic ecosystem.

Exogenous: originating from outside an organism.

Extracellular: outside of a cell, space between cells.

Faculative: not restricted to a particular mode of life.

Flagellum: a locomotory hairlike structure that serves to cause the movement of a cell or a substance, such as moving water through the sponge water canals.

Foramen: a hole, opening on a gemmule through which archaeocytes are released.

Foramina: the plural of foramen, more than one foramen.

Gametes: sperm and eggs, a haploid male or female germ cell which is capable of uniting with germ cell of the opposite sex in sexual reproduction to form a zygote.

Gametogenesis: production of gametes, producing eggs or sperm.

Gemmular theca: the protective case portion of a gemmule, case surrounding a mass of archaeocytes, often composed of three layers: outer membrane, pneumatic layer, and inner layer.

Gemmulation: to produce gemmules.

Gemmule: asexual reproductive structure of sponges produced for surviving hibernation, estivation, and desiccation.

Gemmulosclere: a type of spicule used in the formation of a gemmule, arranged to surround and provide protection to a mass of archeocytes or thymocytes.

Genetic drift: Random fluctuations in the frequency of an allele from one generation to the next in a small population, can result in the loss of a gene from a population's gene pool.

Glycoprotein: a compound composed of a carbohydrate and a protein in which short carbohydrate (oligosaccharide) chains are covalently bonded to a polypeptide side chain, a food source for archaeocytes.

Gonochoristic: to be either male or female; two distinct sexes male and female in the same population reproducing by cross-fertilization.

Haploid: to have a single set of chromosomes (n).

Herbarium: a depository of preserved plants.

Hermaphrodite: the condition of having both male and female reproductive structures in a single individual.

Heterotroph: an organism that obtains organic food molecules by eating other organisms.

Hibernation: a state of dormancy during winter.

Homology: the study of relationships based on the existence of shared or common ancestry, similarity of a structure or function of parts based on descent from a common evolutionary ancestor.

Hydrographic: adjective form of hydrography.

Hydrography: the science of surveying and charting bodies of water.

Hydrophilic: an attraction to water, water loving.

Instar: a developmental stage occurring between molts of insect/arthropods.

Invagination: a cavity or pouch, in sponges not as deep as a chamber.

Invertebrate: animal lacking a vertebral column.

Isotropic: to be uniform in all orientations.

Jurassic Period: the geologic period that extends from 201 million years ago to 145 million years ago; time period from the end of the Triassic to the beginning of the Cretaceous.

Larva: distinct juvenile form of some animals which undergoes metamorphosis to become an adult.

Larvae: plural form of larva.

Lentic waters: stationary or relatively still water, such as a pond or lake.

Leuconoid: from leucon, sponge having choanocyte chambers in its water canal system.

Lipoprotein: a compound composed of a lipid and a protein, used as a food source by archaeocytes.

Lophocytes: mobile cells in sponges which produce collagen.

Lotic: flowing water, such as a river.

Lubomirskiidae: a family of freshwater sponges which is limited in distribution to Lake Baikal in Russia.

Malawispongiidae: a family of freshwater sponges, found in ancient lakes of Malawi and Tanganyika.

Megasclere: largest type of spicule found in the skeleton of a sponge.

Meiosis: cell division that reduces the chromosome number from 2n to n in cells which become gametes, also known as reduction division.

Mesenchyme: the gelatinous matrix within a sponge. It fills the space between the external pinacoderm and the internal choanoderm.
Mesoglea: see mesenchyme.
Mesohyl: see mesenchyme.
Metamorphosis: process resulting in a change of form and function undergone by an animal from embryo to adult.
Metaniidae: a family of freshwater sponges.
Metazoa/Metazoan: a major division of the animal kingdom that comprises all animals other than protozoans and sponges; multicellular animals with differentiated tissues.
Metschnikowiidae: a family of freshwater sponges.
Microbiome: a community of microorganisms consisting of bacteria, fungi, and viruses that exist in a particular environment.
Micron: a unit of length measurement, 1×10^{-6}m, scientific notation is μm.
Microscleres: small spicules scattered throughout the body of some sponge and add structural reinforcement for the sponge body.
Microvilli: very small threadlike or fingerlike projections on a choanocyte which form a very fine mesh capable of filtering food sources from the water.
Mixotrophic: deriving nourishment utilizing both autotrophic and heterotrophic mechanisms.
Morph: form, arrangement, structure.
MYA: notation for million years ago.
Myocytes: muscle-like cells containing actin and myosin filaments, capable of contraction, responsible for opening and closing porocytes.
Neofunctionalization: the process by which a gene acquires a new function after a gene duplication event.
Neuroptera: insect order, net–winged insects.
Obligate: restricted to a particular or specific mode of life.
Occlusion: the blockage or closing of a vessel, tube, or hollow organ.
Oocyte: female gametocyte or germ cell which may undergo meiotic division to form an ovum.
Oscules: Oscula
Oviposit: laying of eggs.
Oxea: spicule pointed at both ends.
Paleozoic Era: time from 541 to 252 million years ago, is subdivided into six geologic periods (from oldest to youngest): the Cambrian, Ordovician, Silurian, Devonian, Carboniferous, and Permian.
Paleospongillidae: a family of extinct freshwater sponges.
Parazoa: division of the animal kingdom that comprises the placozoa and sponges; animals lacking organs and tissues.
Parenchymella: free swimming sponge larvae having a single cell layer of ciliated epithelium and a large vacuole.
Pauci: few.
Paucispicular: having few spicules.

Periphyton: a mixture of algae, heterotrophic microorganisms, and detritus attached to submerged objects.

Peritrophic: a tubular chitinous sheath that lines the midgut of many insects.

Permo-Carboniferous: time period including the latter parts of the Carboniferous and early part of the Permian period.

Phagocyte: cell capable of engulfing and absorbing bacteria and other small particles.

Phagosome: a vacuole in the cytoplasm of a phagocyte, containing a phagocytosed particle enclosed within a part of the cell membrane.

Phenotypic plasticity: the ability of an organism to change its phenotype (observable characteristics) in response to changes in the environment.

Physiographic province: geographic region with a characteristic geomorphology, and subsurface rock type, typically with differences in life-forms when compared to other physiographic provinces.

Phytoplankton: photosynthetic microscopic plantlike organisms forming the base of the marine food web.

Pinacoderm: outer most layer of sponge cells, equivalent to the epidermis of higher animals.

Pinacocytes: type of cell found in the pinacoderm, an epidermal–like cell.

Pleistocene: geological epoch which lasted from about 2,588,000 years ago to 12,000 years ago, spanning the world's recent period of repeated glaciations.

Pneumatic layer: middle layer of a three–layer gemmular theca, consists of spongin that is porous with chambers or meshes.

Porifera: phylum that comprises the sponges.

Porocytes: tubular cells found within the pinacoderm which make up the pores of a sponge known as ostia.

Potamolepidae: one of the families of freshwater sponges.

Precambrian: period that covers the vast bulk of the Earth's history, beginning with the planet's creation about 4.5 billion years ago and ending with the emergence of complex, multicellular lifeforms nearly four billion years later.

Predation: the act of an animal killing and eating another animal.

Primary Production: synthesis of organic compounds from carbon dioxide. It primarily occurs through photosynthesis.

Propagules: structures that can give rise to a new individual organism.

Prosopyle: pore or opening through water passes upon entry into a choanocyte chamber.

Pseudobirotulate: spicule with a shaft having hooks or spines on the ends.

Pupation: transformation life stage occurring between larva and adult stage.

Quiescence: a state of dormancy, delayed development controlled by exogenous factors.

Radial: arranged like rays from a central point, or the radii of a circle.

Rhagonoid: from rhagon, a cone-shaped larval body stage which develops into a leuconoid body.

Sclerocytes: sponge cells which secrete calcareous or siliceous spicules.

Secondary production: the net quantity of energy stored in the tissues of heterotrophs.

Sexual reproduction: form of reproduction where two morphologically distinct haploid (n) gametes, egg and sperm, fuse together resulting in a diploid (2n) zygote.
Siphon: tube through which water flows.
Somatic cells: cells not involved in reproduction, any cell not a gamete.
Spermatocytes: cells which become sperm.
Sphaerocladina: an order of marine sponges, the sister taxon to freshwater sponges.
Spicule: minute structure composed of silica dioxide (in freshwater sponges) which make up a part of the skeleton and provide structural support to a freshwater sponge.
Spongillida: taxonomic order of freshwater sponges.
Spongillidae: one of the families of freshwater sponges.
Spongin: modified type of collagen protein, forms the fibrous skeleton, gives sponges flexibility.
Strongyle: spicule having rounded ends.
Syconoid: sponge having radial canals lined with choanocytes in its water canal system.
Taxon: taxonomic group of any rank, such as Phylum, Class, Order, Family, Genus.
Taxonomy: field of biology which defines groups of organisms based on shared characteristics and giving names to those groups.
Theca: capsule or structure serving as a protective covering surrounding an organ or organism; in sponges protecting staminal cells.
Thesocyte; a cell derived from an archaeocyte which contains reserved food in form of yolk platelets.
Totipotent: ability of a cell to change its structure and function to become a different cell type; a stem cell.
Trabecula: structure that separates or divides.
Trichoptera: taxonomic order of caddisflies.
Umbonate: rounded protuberance.
Viviparous: giving birth to live young.
Zooplankton: heterotrophic plankton
Zygote: diploid (2n) cell resulting from the fusion of two haploid (n) gametes.

SELECTED REFERENCES

Adamson, D. A., J. D. Clark, and M. A. J. Williams. 1987. Pottery tempered with sponge from the White Nile, Sudan. African Archaeological Review 5:115–27.

Ahmed, S. F., P. S. Kumar, M. Kabir, F. T. Zuhara, A. Mehjabin, N. Tasannum, A. T. Hoang, Z. Kabir, and M. Mofijur. 2022. Threats, challenges and sustainable conservation strategies for freshwater biodiversity. Environmental Research 214 (Pt 1):113808. doi: 10.1016/j.envres.2022.113808.

Alison, C. C., T. M. Frost, and J. M. Fischer. 1999. Sponge distribution and lake chemistry in northern Wisconsin lakes: Minna Jewell's survey revisited. Memoirs of the Queensland Museum 44:93–99.

Annandale, N. 1910. Fresh-water sponges in the collection of the United States Museum. Part IV. Notes on the fresh-water sponge, *Ephydatia japonica*, and its allies. Proceedings United States National Museum 38:183.

———. 1911. Freshwater sponges, hydroids and polyzoa. Porifera. In: Shipley, A. D. (Ed). Fauna of British India, including Ceylon and Burma. Taylor and Francis, London. p. 27–126, 241–45.

Annesley, J., J. Jones, and D. Watermolen. 2008. Wisconsin freshwater sponge species documented by scanning electron microscopy. Journal of Freshwater Ecology 23:263–72.

Asadzadeh, S. S., P. S. Larsen, H. U. Riisgård, and J. H. Walther. 2019. Hydrodynamics of the leucon sponge pump. Journal Royal Society Interface 16(150): 20180630.

Ashley, J. M. 1913. Fresh water sponges of Illinois and Michigan. Master Thesis. University of Illinois. pp 1–30.

Bader, R. B. 1984. Factors affecting the distribution of a freshwater sponge. Freshwater Invertebrate Biology 3:86–95.

Barbeau, M. A., H. M. Reiswig, and L. C. Rath. 1982. Hatching of freshwater sponge gemmules after low temperature exposure: *Ephydatia muelleri* (Porifera: Spongillidae). Journal of Thermal Biology 14:225–31.

Barnes, D. K., and T. E. Lauer. 2003. Distribution of freshwater sponges and bryozoans in northwest Indiana. Proceedings of the Indiana Academy of Science 112:29–35.

Barnes, R. D. 1987. Invertebrate Zoology, 5th ed. Saunders College Publishing, Philadelphia, PA. p. 1–1089.

Barton, S. H., and J. S. Addis. 1997. Freshwater sponges (Porifera: Spongillidae) of western Montana. Great Basin Naturalist 57:93–103.

Bass, D., and C. Volkmer–Ribeiro. 1998. *Radiospongilla crateriformis* (Porifera, Spongillidae) in the West Indies and taxonomic notes. Iheringia, Série Zoologie Porto Alegre 85:123–28.

Benfey, T. J., and H. M. Reisweig. 1982. Temperature, pH, and photoperiod effects upon gemmule hatching in the freshwater sponge, *Ephydatia muelleri* (porifera, Spongillidae). Journal of Experimental Zoology 221:13–21.

Bond, C., and A. K. Harris. 1988. Locomotion of sponges and its physical mechanism. Journal of Experimental Zoology 246:271–84.

Boury-Esnault, N., and N. K. Rützler. 1997. Thesaurus of sponge morphology. Smithsonian Contributions to Zoology Number 596, Smithsonian Institution Press, Washington, DC.

Bowerbank, J. S. 1863. A Monograph of the Spongillidae. *Proceedings of the Zoological Society of London* 1863:440–72, pl. 38.

Burton, M. 1949. Observations on littoral sponges, including the supposed swarming of larvae, movement, and coalescence in mature individuals, longevity and death. Proceedings Zoological Society of London 118:893–915.

Carballo J. L., and J. J. Bell. 2017. Chapter 1: Climate change and sponges: an introduction. Pp 1–11 In: Carballo, J., and J. Bell (Eds.). Climate Change, Ocean Acidification and Sponges. Springer, Cham, Switzerland.

Carballo, J. L., J. A. Cruz-Barraza, B. Yáñez, and P. Gómez. 2018. Taxonomy and molecular systematic position of freshwater genus *Racekiela* (Porifera: Spongillida) with the description of a new species from North-west Mexico. Systematics and Biodiversity 16(2):160–70.

Carballo, J. L., P. Gómez, J. A. Cruz–Barraza, and B. Yáñez. 2021. Taxonomy and molecular systematic position of the freshwater genus *Heteromeyenia* (Porifera: Spongillida) with the description of a new species from Mexico, Systematics and Biodiversity 19:940–56.

Cathers, A. 2019. Determining the provenance of freshwater sponge spicule inclusions in Pre-Columbian Amazonian ceramics. Presented at the 84th annual meeting of The Society for American Archaeology. Albuquerque, NM.

Causey, D. 1951. Freshwater sponges of Arkansas. Journal of the Arkansas Academy of Science 4:89–90.

Cheatum, E. P., and J. P. Harris. 1953. Ecological observations upon the fresh-water sponges in Dallas County, Texas. Field and Laboratory 31: 97–103.

Chernogor, L., E. Klimenko, I. Khanaev, and S. Belikov. 2020, Microbiome analysis of healthy and diseased sponges *Lubomirskia baicalensis* by using cell cultures of primmorphs. DOI10.7717/peerj.9080.

Copeland, J. E., S. C. Kunigelis, E. A. Stuart, and K. A. Hanson. 2020. First records of freshwater sponges (Porifera: Spongillidae) for Great Smoky Mountain National Park. Journal of the Tennessee Academy of Science 95(1): 59–62.

Copeland, J., S. Kunigelis, J. Tussing, T. Jett, and C. Rich. 2019. Freshwater sponges (Porifera: Spongillida) of Tennessee. The American Midland Naturalist 181:310–26.

Copeland, J., R. Pronzato and R. Manconi. 2015. Discovery of living Potamolepidae (Porifera: Spongillina) from Nearctic freshwater with description of a new genus. Zootaxa 3957:37– 48.

Copeland, J. E., J. A. Tussing, T. M. Jett, and S. Kunigelis. 2014. Freshwater Porifera of eastern Tennessee. Journal Tennessee Academy of Science 90:23.

Corallini, C., and E. Gaino. 2003. The Caddisfly, *Ceraclea fulva* and the freshwater sponge *Ephydatia fluviatilis*: a successful relationship. Tissue and Cell 35:1–7.

Costa, M. L., D. C. Kern, A. H. E. Pinto, and J. R. T. Souza. 2004. The ceramic artifacts in archaeological black earth (terra preta) from lower Amazon region, Brazil: mineralogy. Acta Amazonica 34:165–78.

Cruz, A. A. V., V. M. Alencar, C. Volkmer-Ribeiro, V. L. Gattas, and E. Luna. 2013. Dangerous waters: outbreak of eye lesions caused by freshwater sponge spicules. Eye 27:398–402.

Dawson, L. H. 1966. Notes on the occurrence of Porifera in Nebraska. Transaction of the Kansas Academy of Science 69(1):96–98.

———. 1932. The marine and freshwater sponges of California. Proceedings United States National Museum 81 (4):1–140.

de Laubenfels, M. W. 1935. A freshwater sponge from southern California. Science 81:154.

Denikina, N. N., E. V. Dzyuba, N. L. Bel'kova, I. V. Khanaey, S. I. Feranchuk, M. M. MaKarov, N. G. Granin, and S. I. Belikov. 2016. The first case of disease of the sponge *Lubomirskia baicalensis*: investigation of its microbiome. Biology Bulletin 43:263–70.

de Ronde, C. E. J., P. Stoffers, D. Garbe-Schonberg, B. Jones, R. Manconi, B. W. Christenson, P. R. L. Browne, et al. 2002. Discovery of active hydrothermal venting in Lake Taupo, New Zealand. Journal Volcan Geothermal Research 15:257–75.

De Santo, E. M., and P. E. Fell. 1996. Distribution and ecology of freshwater sponges in Connecticut. Hydrobiologia 341:81–89.

Dominey, W. 1987. Sponge-eating by *Pungu maclareni*, an endemic Cichlid fish from Lake Barombi Mbo, Cameroon. National Geographic Research 3:389–93.

Dudgeon, D., A. H. Arthington, M. O. Gessner, Z. I. Kawabata, D. J. Knowler, C. Lévêque, R. J. Naimen, et al. 2006. Freshwater biodiversity: importance, threats, status and conservation challenges. Biological Reviews 81:143–82.

Dunagan, S. P. 1999. A North American freshwater sponge (*Eospongilla morrisonensis* new genus and species) from the Morrison Formation (Upper Jurassic), Colorado. Journal of Paleontology 73: 389–93.

Duncan, T. O. 1977. Freshwater sponges in Beaver Reservoir, Northwest Arkansas. The Southwestern Naturalist 22:140.

Ereskovsky, A. L. 2004. Comparative embryology of sponges and its application for poriferan phylogeny. Bollettino dei musei e degli istituti biologici dell'Universita di Genova 68:301–18.

Eshleman III, S. K. 1950. A key to Florida's freshwater sponges, with descriptive notes. Quarterly Journal Florida Academy of Science 12:35–44.

Etnier, D. A., and W. C. Starnes. 1993. The fishes of Tennessee. The University of Tennessee Press, Knoxville, Tennessee. p. 1–681.

Evans, K. L., and C. L. Kitting. 2010. Documentation and identification of the one known freshwater sponge discovered in the California Delta. The Open Marine Biological Journal 4:82–86.

Evans, K. L., and D. J. S. Montagnes. 2019. Freshwater sponge (Porifera: Spongillidae) distribution across a landscape: environmental tolerances, habitats, and morphological variation. Invertebrate Zoology, 138(3).

Evans, R. 1899. The structure and metamorphosis of Spongilla lacustris. Journal of Cell Sciences 2–42 (168):363–476.

Fell, P. E. 1990. Environmental factors affecting dormancy in the freshwater sponge *Eunapius fragilis* (Leidy). Invertebrate Reproduction and Development 18:213–19.

Fell, P. E., and L. J. Bazer. 1990. Survival of the gemmules of *Anheteromeyenia ryderi* (Potts) following aerial exposure during winter in New England. Hydrobiologia 190:241–46.

Fishelson, L. 1981. Observations on the moving colonies of the genus *Tethya* (Demospongia, Porifera). Zoomorphology 98:89–99.

———. 1980. Selection in sponge feeding processes. In: Smith, D. C., and Y. Tiffon (Eds.). Nutrition in lower metazoa. Pergammon Press, Oxford, England. p. 33–44.

Frost, T. M. 1991. Porifera. In: Thorp, J. H., and A. P. Covich (Eds.). Ecology and Classification of North American Freshwater Invertebrates. Academic Press, San Diego, California. p. 95–124.

Frost, T. M., G. S. de Nagy, and J. J. Gilbert. 1982. Population dynamics and standing biomass of the freshwater sponge, *Spongilla lacustris*. Ecology 63:1203–10.

Frost, T. M., L. E. Graham, J. E. Elias, M. J. Haase, D. W. Kretchmer and D. W. Kranzfelder. 1997. A yellow-green algal symbiont in the freshwater sponge *Corvomeyenia everetti*: convergent evolution of symbiotic associations. Freshwater Biology 38: 396–99.

Frost, T. M., and C. E. Williamson. 1980. *In situ* determination of the effect of symbiotic algae on the growth of the fresh-water sponge *Spongilla lacustris*. Ecology 61: 1361–70.

Gaikwad, S., Y. S. Shouche, and W. N. Gade. 2016. Microbial community structure of two freshwater sponges using Illumina MiSeq sequencing revealed high microbial diversity. AMB Express 6:40–49.

Gee, N. G. 1931. Fresh-water sponges from Australia and New Zealand. Records of the Australian Museum 18(2): 25–62.

———. 1931. North American fresh-water sponges. Science 73:501–2.

———. 1934. Some new records of occurrence of North American fresh-water sponges. Science 80:248.

Gilbert, J. J. 1975. Field experiments on gemmulation in the freshwater sponge *Spongilla lacustris*. Transactions American Microscopical Society 96:62–67.

Gilbert, J. J., and T. L. Simpson. 1976. Sex reversal in a freshwater sponge. Journal of Experimental Zoology 195:145–51.

Gillis, P. L., and G. L. Mackie. 1994. Impact of the zebra mussel, *Dreissena polymorpha*, on populations of Unionidae (Bivalvia) in Lake St. Clair. Canadian Journal of Zoology 72:1260–71.

Gold, D. A., J. Grabenstatter, A. de Mendoza, A. Riesgo, I. Ruiz-Trillo, and R. E. Summons. 2016. Sterol and genomic analyses validate the sponge biomarker hypothesis. Proceedings of the National Academy of Sciences of the United States of America 113 (10) 2684–89.

Gómez, P., J. L. Carballo, J. A. Cruz-Barraza, and M. Camacho-Cancino. 2019. On the genus *Racekiela* in Mexico: molecular and morphological description of *Racekiela cresciscrystae* n. sp. Journal of Natural History 53: 1351–68.

Gost, M., S. Pinya, A. Sureda, S. Tejada, and P. Ferriol. 2023. Effect of alkalinity and light intensity on the growth of the freshwater sponge *Ephydatia fluviatilis* (Porifera: Spongillidae). Aquatic Ecology 57:353–67.

Gugel, J. 1996. The sponge mite *Unionicola minor* (Acari, Hydrachnellae, Unioncolidae) in a pond at the spring of the Darmbach near Darmstadt. Hessische Faunistische Briefe 15 (1): 11–16.

———. 2001. Lifecycle and ecological interactions of freshwater sponges (Porifera: Spongillidae). Limnologica 31(3):159–98.

Harrison, F. W. 1971. A taxonomic investigation of the genus *Corvomeyenia* Weltner (Spongillidae) with an introduction to *Corvomeyenia carolinensis sp. nov.* Hydrobiologia 38: 123–40.

———. 1974. Porifera. In: Hart Jr., C. W., and S. L. H. Fuller (Eds.). Pollution ecology of freshwater invertebrates. Academic press, New York. p. 29–66.

———. 1977. The taxonomic and ecological status of the environmentally restricted spongillid species of North America. III *Corvomeyenia carolinensis* Harrison, 1971. Hydrobiology 56: 187–90.

———. 1979. The taxonomic and ecological status of the environmentally restricted spongillid species of North America. V. *Ephydatia subtilis* (Weltner) and *Stratospongilla penneyi* sp. nov. Hydrobiologia 65:99–105.

———. 1988. Methods in Quaternary Ecology #4. Freshwater sponges. Geoscience Canada: Journal of the Geological Association of Canada 15:193–98.

Harsha, R. E., J. C. Francis, and M. A. Poirrier.1983. Water temperature: a factor in the seasonality of two freshwater sponge species, *Ephydatia fluviatilis* and *Spongilla alba*. Hydrobiology 102:145–50.

Hoff, C. C. 1943. Some records of sponges, branchiobdellids, and molluscs from the Reelfoot Lake region. Journal of Tennessee Academy of Science 18:223–27.

Hooper, J. N. A., R. W. M. van Soest, and A. Pisera. 2011. Phylum Porifera Grant, 1826. In: Z. Q. Zanh (Ed.). Animal biodiversity: an outline of higher-level classification and survey of taxonomic richness. Zootaxa 3148:1–237.

Isom, B. G. 1968 New distribution records for aquatic neuropterans, Sisyridae (Spongilla-flies) in the Tennessee River drainage. Journal of the Tennessee Academy of Science 43:109–10.

Jewell, M. E. 1935. An ecological study of the freshwater sponges of northeastern Wisconsin. Ecological Monograph 5:461–504.

———. 1939. An ecological study of the freshwater sponges of Wisconsin, II. The influence of calcium. Ecology 20:11–28.

Jewell, M. E., and H. W. Brown. 1929. Studies on northern Michigan bog lakes. Ecology 10:427–75.

John, P. D., A. E. Bogan, K. M. Burkhead, J. R. Cordeiro, J. T. Garner, P. H. Hatfield, D. A. W. Lepitzki, *et al.* 2013. Conservation status of freshwater gastropods of Canada and the United States. Fisheries 38:247–82.

Jones, M. L., and K. Rützler. 1975. Invertebrates of the upper chamber, Gatún Locks, Panama Canal, with emphasis on *Trochospongilla leidii* (Porifera). Marine Biology 33:57–66.

Kamaltynov, R. M., V. I. Chernykh, Z. V. Slugina, and E. B. Karabanov. 1993. The consortium of the sponge *Lubomirskia baikalensis* in Lake Baikal, East Siberia. Hydrobiologia 27:179–89.

Kernodle, J. M. 1972. Tributary river basins in Tennessee. Tennessee Div. Water Resources (Misc. Pub. 8). 5 p.

Kintner, E. 1939. Notes on Indiana fresh-water sponges. Proceedings Indiana Academy of Science 48:244–45.

Kondrashov, F. A. 2012. Genomic duplication as a mechanism of genomic adaptation to a changing environment. Proceedings of the Royal Society 279:5048–57.

Kunigelis, S. C., and J. E. Copeland. 2014. Identification of Isolated and *in situ* freshwater sponge spicules of eastern Tennessee. Microscopy and Microanalysis 20 (Suppl. 3):1294–95.

Larsen, P. S, and H. U. Riisgård. 1994. The sponge pump. Journal of Theoretical Biology 168:53–63.

Lauer, T. E. 2003. Distribution of freshwater sponges and bryozoans in Northwest Indiana. Proceedings of the Indiana Academy of Science 112:29–35.

Lauer, T. E., and A. Spacie. 1996. New records of freshwater sponges (Porifera) for southern Lake Michigan. Journal Great Lakes Research 22:77–82.

———. 2000. The effects of sponge (Porifera) biofouling on zebra mussel (*Dreissena polymorpha*) fitness: reduction of glycogen, tissue loss, and mortality. Journal of Freshwater Ecology 15:83–92.

———. 2004. An association between freshwater sponges and the zebra mussel in a southern Lake Michigan harbor. Journal of Freshwater Ecology 19: 631–37.

——— 2004. Space as a limiting resource in freshwater systems: competition between zebra mussels (*Dreissena polymorpha*) and freshwater sponges (Porifera). Hydrobiologia 517:137–45.

Lauer, T. E., A. Spacie, and D. K. Barnes. 2001. The distribution and habitat preference of freshwater sponges (Porifera) in four southern Lake Michigan harbors. The American Midland Naturalist 146:243–53.

Lehmkuhl, D. M. 1970. A North American Trichoptera larva which feeds on freshwater sponges (Trichoptera: Leptoceridae; Porifera: Spongillidae). American Midland Naturalist 84:278–80.

Lemonie, N., N. Buell, A. Hill, and M. Hill. 2007. Assessing the utility of sponge microbial Symbiont communities as models to study climate change: a case study with *Halichondria bowerbanki*. In: Custódio, M. R. (Ed.). Porifera research: biodiversity, innovation and sustainability. p 419–25.

Leys, S. P., G. Yahel, A. Reidenbach, V. Tunnicliffe, U. Shavit, and H. Reiswig. 2011. The Sponge pump: the role of current induced flow in the design of the sponge body plan. PloS ONE:6: e27787. https://doi.org/10.1371/journal.pone.0027787.

Lindenschmidt, M. J. 1950. A new species of freshwater sponge. Transactions American Microscopical Society 69(2):214–16.

Luter, H. M., and N. S. Webster. 2017. Chapter 9: Sponge disease and climate change. In: Carballo, J., and J. Bell (Eds.). Climate Change, Ocean Acidification and Sponges. Springer, Cham, Switzerland. p 421–28.

Luther, E. T. 1977. Our restless earth: geologic regions of Tennessee. The University of Tennessee Press, Knoxville, TN. p. 1–106.

Mackie, G. L. 1991. Biology of exotic zebra mussel, *Dreissena polymorpha*, in relation to native bivalves and its potential impact in Lake St. Clair. Hydrobiologia 219:251–68.

Magnusson, T. 2005. A comparison of microbial communities associated with the freshwater sponges *Radiospongilla crateriformis* and *Eunapius fragilis*. M. S. Thesis. The University of Alabama at Birmingham.

Mah, J. L., K. K. Christensen-Dalsgaard, and S. P. Keys. 2014. Choanoflagellate and choanocyte collar-flagellar systems and the assumption of homology. Evolution and Development 16:25–37.

Manconi, R., J. Copeland, S. Kunigelis, and R. Pronzato. 2022. Biodiversity of Nearctic inland Water: discovery of the genus *Heterorotula* (Porifera, Spongillida, Spongillidae) in the Appalachian Mountains, with biogeographical implications and description of new species. ZooKeys 1110:103–20.

Manconi, R., and R. Pronzato. 2000. Rediscovery of the type material of *Spongilla lacustris* (L., 1759) in the Linnean Herbarium. Italian Journal of Zoology 67:89–92.

———. 2002. Suborder Spongillina subord. nov.: Freshwater Sponges. In: Hooper, J. N. A., and R. W. M Van Soest (Eds.). Systema Porifera: A Guide to the Classification of Sponges. Vol. 1. Kluwer Academic/Plenum Publishers, New York. p. 921–1019.

———. 2007. Gemmules as a key structure for the adaptive radiation of freshwater sponges: a morpho-functional and biogeographical study. Porifera Research: Biodiversity, Innovation and Sustainability. p. 61–77.

———. 2008. Global diversity of sponges (Porifera: Spongillina) in freshwater. Hydrobiologia 595:27–33.

———. 2015. Chapter 8: Phylum Porifera, 133–57. In: Thorp, J., and D. C. Rogers (Eds.). Thorp and Covich's Freshwater Invertebrates, 4th ed. Vol. 1, Ecology and General Biology. Academic Press, London, UK.

———. 2016. Chapter 3: Phylum Porifera, 39–83. In: Thorp, J., and D. C. Rogers (Eds.). Thorp and Covich's Freshwater Invertebrates, 4th ed. Vol. 2, Keys to Nearctic Fauna. Academic Press, London, UK.

Manconi, R., and R. Pronzato. 2016. How to survive and persist in temporary freshwater? Adaptive raits of sponges (Porifera: Spongillidae): a review. Hydrobiologia 782:11–22.

Manconi, R., N. Ruengsawang, V. Vannachak, C. Hanjavant, N. Sangpradub, and R. Pronzato. 2013. Biodiversity in South East Asia: an overview of freshwater sponges (Porifera: Demospongiae: Spongillina). Journal Limnology 72:313–26.

Mann, K. H., R. H. Kowalczewski, A. Kowalczewski, T. J. Lack, C. P. Mathews, and L. McDonald. 1972. Productivity and energy flow at all trophic levels in the River Thames, England. In: Kajak, Z., and A. Hillbricht-Ilkowska (Ed). Productivity problems of Freshwaters. PWN, Polish Scientific Publishers, Warsaw, Poland. p. 579–96.

Matsuoka, T, and Y. Masuda. 2000. A new potamolepid freshwater sponge (Demospongiae) from the Miocene Nakamura Formation, central Japan. Paleontological Research 4:131–37.

McAuley, D. G., and J. R. Longcore. 1988. Foods of juvenile ring-neck ducks: relationship to wetland pH. Journal of Wildlife Management 52:177–85.

Mobley, A. S. 2010. The bacterial community of a freshwater sponge, *Radiospongilla cerebellata*: a comparison of terminal restriction fragment length polymorphisms (T-RFLP) and 16S RRNA clone library methods. M. S. Thesis. The University of Alabama at Birmingham. p. 63

Montana Natural Heritage Program. 2020. *Heteromeyenia baileyi*. Montana Field Guide. http://FieldGuide.mt.gov/speciesDetail.aspx?elcode=IZSPN05060.

Moore, W. G 1951. Louisiana fresh-water sponges with observations on the ecology of the sponges of the New Orleans area. Bulletin Ecological Society of America 32:63.

———. 1953. Notes on Louisiana Spongillidae. Proceedings Louisiana Academy of Science 16:42–43.

Moulton, J. K. 2007. New additions to the caddisfly fauna (Trichoptera) of Tennessee and Virginia. Entomological News 118:209–10.

NatureServe Network. 2020. Species richness for imperiled aquatic invertebrates. From the Map of Biodiversity Importance Initiative: A collaborative effort to identify the places most important for conserving at-risk species. Arlington, Virginia, USA. Available at https://www.natureserve.org/node/1138. Additional biodiversity conservation data and expertise are available by contacting the NatureServe Network. NatureServe Network contacts can be found here: https://www.natureserve.org/ns-network-directory.

Neel, J. K., and R. L. Post 1983. Observations of North Dakota Sponges (Haplosclerina: Spongillidae) and Sisyrids (Neuroptera: Sisyridae). The Great Lakes Entomologist 16 (4):1–6.

Neidhoefer, J. R. 1938. *Carterius tenosperma* Potts, a species of fresh-water sponge new to Wisconsin. Transactions of the American Microscopical Society 57:82–84.

———. 1940. The fresh–water sponges of Wisconsin. Transactions Wisconsin Academy of Sciences, Arts, and Letters 32:177–79.

Nichols, H. T., and T. H. Bonner. 2014. First record and habitat associations of *Spongilla cenota* (Class Demospongiae) within streams of the Edward Plateau, Texas, USA. The Southwestern Naturalist 59: 467–72.

Nielsen, J. G., W. Schwarzhans, and D. M. Cohen. 2012. Revision of *Hastatobythites* and *Saccogaster* (Teleostei, Bythitidae) with three new species and a new genus. *Zootaxa* 3579: 1–36.

Oficjalski, P. 1937. Spongia fluviatilis (Badiaga), *Pharmazeutische Zentralhalle fur Deutschland* 78:173–75.

Økland, K. A., and J. Økland. Freshwater sponges (Porifera: Spongillidae) of Norway distribution and ecology. Hydrobiologica 330: 1–30.

Old, M. C. 1932. Contribution to the biology of fresh-water sponges (Spongillidae). Michigan Academy Science, Arts, and Letters 17:663–79.

———. 1932. Taxonomy and distribution of the fresh-water sponges (Spongillidae) of Michigan. Pap. Mich. Acad. Science, Arts, and Letters 15:439–77.

———. 1932. Environmental selection of the fresh-water sponges (Spongillidae) of Michigan. Transactions American Microscopy Society 51:129–36.

———. 1932. Delaware freshwater sponges. Transactions American Microscopical Society 51:239–42.

———. 1936. Additional North American freshwater sponge records. Transactions American Microscopical Society 55:11–13.

Paduano, G. M., and P. E. Fell. 1997. Spatial and temporal distribution of freshwater sponges in Connecticut lakes based upon analysis of siliceous spicules in dated sediment cores. Hydrobiologia 350:105–21.

Palmer, S. R. 2008. Eco logic . . . from the Nature Conservancy--Averting a water supply crisis while protecting endangered species: partnerships pay off for Tennessee's Duck River. Journal American Water Works Association 100 (8): 40–43.

Parchment, J. G. 1966. Notes on the ecology of sponges. Journal of the Tennessee Academy of Science 41:65.

Parfenova, V. V., I. A. Terkina, T. Ya Kostornova, I. G. Nikulina. V. I. Chernykh, and E. A. Maksimova. 2008. Microbial community of freshwater sponges in Lake Baikal. Biology Bulletin 35:374–79.

Parfin, S.I., and A. B. Gurney. 1956. The spongilla-flies, with special reference to those of the western hemisphere (Sisyridae, Neuroptera). Proceedings of the United States National Museum 105:421–529.

Parmalee, P. W., and A. E. Bogan. 1998. The Freshwater Mussels of Tennessee. The University of Tennessee Press, Knoxville, TN. p. 1–328.

Paulus, W., and N. Weissenfels, 1986. The spermatogenesis of *Ephydatia fluviatilis* (Porifera). Zoomorphology 106:155–62.

Pejin, B., A. Talevski, A. Ciric, J. Glamoclija, M. Nikolic, T. Talevski, and M. Sokovic. 2014. *In vitro* evaluation of antimicrobial activity of the freshwater sponge *Ochridaspongia rotunda* (Arndt, 1937). Natural Products Research 28:1489–94.

Penny, J. T. 1931. Notes on fresh-water sponges and their epithelial membranes. Journal Elisha Mitchell Scientific Society 46:240–46.

———. 1960. Distribution and bibliography (1892–1957) of the freshwater sponges. University of South Carolina Publications 3 (1):1–97.

Penny, J. T., and A. A. Racek. 1968. Comprehensive revision of a worldwide collection of freshwater sponges (Porifera: Spongillidae). Smithsonian Institution Press, Washington, D.C. p. 1–184.

Perkins, D. M., J. Reiss, G. Yvon–Durocher, and G. Woodward. 2010. Global change and food webs in running waters. Hydrobiologia 657:181–98.

Pisera, A. 2004. What can we learn about siliceous sponges from paleontology, In: Pansini, M., R. Pronzato, G. Bavestrello, R. Manconi (Eds.). Sponge Science in the New Millennium. Bollettino dei Musei e degli Instuti Biologici dell' Università di Genova 68:55–69.

Pisera, A., and A. Saez. 2003. Paleoenvironmental significance of a new species of freshwater sponge from the late Miocene Quillagua Formation (N Chile). Journal of South American Earth Studies 15:847–52.

Pisera, A., P. A. Siver, and A. P. Wolf. 2013. A first account of freshwater Potamolepid sponges (Demospongiae, Spongillina, Potamolepidae) from the middle Eocene: biogeographic and paleoclimatic implications. Journal of Paleontology, 87:373–78.

Poirrier, M. A. 1969. Louisiana fresh-water sponges: taxonomy, ecology, and distribution. PhD Thesis. Louisiana State University, Baton Rouge, LA. p. 1–173.

———. 1969. Some fresh-water sponge hosts of Louisiana and Texas spongilla-flies, with new locality records. American Midland Naturalist 81:573–75.

———. 1972. Additional records of Texas freshwater sponges (Spongillidae) with the First record of *Radiospongilla cerebellata* (Bowerbank, 1863) from the Western Hemisphere. The Southwestern Naturalist 16:434–35.

———. 1974. Ecomorphic variation in gemmuloscleres of *Ephydatia fluviatilis* Linnaeus (Porifera: Spongillidae) with comments upon its systematics and ecology. Hydrobiologia 44:337–47.

———. 1976. A taxonomic study of *Spongilla alba, S. cenota, S. wagneri* study group (Porifera: Spongillidae) with ecological observations of *S. alba*. In: Harrison, F. W., and R. R. Cowden (Eds.). Aspects of Sponge Biology, Academic Press, New York. p. 203–13.

———. 1977. Systematic and ecological studies of *Anheteromeyenia ryderi* (Porifera, Spongillidae) in Louisiana. Transactions American Microscopy Society 96:62–67.

———. 1978. *Corvospongilla becki* n. sp., A new fresh-water sponge from Louisiana. Transactions American Microscopy Society 97:240–43.

———. 1990. Freshwater sponges (Porifera: Spongillidae) from Panama. Hydrobiologia 194:203–6.

Poirrier, M. A., and Y. M. Arceneaux. 1972. Studies on southern Sisyridae (spongilla-flies) with a key to the third-instar larvae and additional sponge-host records. The American Midland Naturalist 88:455–58.

Poirrier, M. A., P. S. Martin, and R. J. Baerwald. 1987. Comparative morphology of microsclere structure in *Spongilla alba, S. cenota*, and *S. lacustris* (Porifera: Spongillidae). Transactions American Microscopical Society 106:302–10.

Poirrier., M. A., and E. F. Strobel. 1969. Fresh-water sponges (Demospongiae: Spongillidae) from Mississippi. The Journal of the Mississippi Academy of Sciences 14:130–31.

Potts, E. 1887. Contributions towards a synopsis of the American forms of freshwater sponges with descriptions of those named by other authors from all parts of the world. Proceedings of the Academy of Natural Sciences of Philadelphia 39:157–279.

———. 1918. Chapter 10, The Sponges (Porifera). In: Ward, H. B., and G. H. Whipple. Freshwater Biology. John Wiley & Sons, New York. p. 301–15.

Proctor, H. C., and G. Pritchard. 1990. Variability in the life history of *Unionicola crassipes*, a sponge-associated water mite (Acari:Unionicolidae). Canadian Journal of Zoology 68:1227–32.

Pronzato, R., and R. Manconi. 1991. Colonization, life cycle and competition in a freshwater sponge association. In: Reitner, J., and H. Keupp (Eds.). Fossil and Recent Sponges. Springer: Berlin. p. 432–44.

———. 1993. Life history of *Ephydatia fluviatilis*: a model for adaptive strategies in discontinuous habitats, In: van Soest, R. W. M., van Kempen, Th. M. G. and Braekman, J. C. (Eds.) Sponges in time and space. Rotterdam: Balkema. p. 327–31.

——— 1994. Adaptive strategies of sponges in inland waters. Italian Journal of Zoology 61:395–401.

Pronzato, R., R. Manconi, and G. Corriero. 1993. Biorhythm and environmental control in the life history of *Ephydatia fluviatilis* (Demospongiae, Spongillidae). Italian Journal of Zoology 60:63–67.

Pronzato R., A. Pisera, and R. Manconi. 2017. Fossil freshwater sponges: taxonomy, geographic distribution, and critical review. Acta Palaeontologica Polonica 62:467–95.

Racek, A. A., and F. W. Harrison. 1975. The systematic and phylogenetic position of *Paleospongilla chubutensis* (Porifera: Spongillidae). Proceedings Linnean Society 99:157–65.

Rader, R. B. 1984. Factors affecting the distribution of a freshwater sponge. Freshwater Invertebrate Biology 3:86–97.

Rader, R. B., and R. N. Winget. 1985. Seasonal growth rate and population dynamics of a freshwater sponge. Hydrobiologia 123:171–76.

Rainbolt, M. L. 1954. Some fresh water sponges of Oklahoma. Proceedings of the Oklahoma Academy of Sciences for 1954. p. 84–87.

Randal, J. E., and W. D. Hartman 1968. Sponge-feeding fishes of the West Indies. Marine Biology 1:216–25.

Rasbold, G. G., L. Calheira, L. Domingos-Luz, L. C. R. Pessenda, U. Pinheiro, and M. M. McGlue. 2023. A Morphological guide of neotropical freshwater sponge spicules for paleolimnological studies. Frontiers of Ecological Evolution 18:1–19.

Rasbold, G. G., U. Pinheiro, L. Domingos-Luz, J. Dilworth, J. P. Thigpen, L. C. S. Pessenda, and M. M. McGlue. 2022. First evidence of an extant freshwater sponge fauna in Jackson Lake, Grand Teton National Park, Wyoming (USA). Inland Waters 12:407–17.

Reisser, W., and M. Widowski. 1992. Endosymbiotic associations of algae with freshwater protozoa and invertebrates. In: Reisser, W. (Ed.). Algae and symbiosis: plants, animals, fungi, viruses, interactions explored. Biopress Limited, Bristol, UK. p.1–19.

Reiswig, H. M., T. M. Frost, and A. Ricciardi. 2010. Porifera, in: Thorp, J. H., and A. P. Covich (Eds.). Ecology and Classification of North American Freshwater Invertebrates. Academic Press, London, UK. p. 91–123.

Reiswig, H. M., and T. L. Miller. 1998. Freshwater sponge gemmules survive months of anoxia. Invertebrate Biology 117:1–8.

Resch, V. H. 1976. Life cycles of invertebrate predators of freshwater sponge. p. 299–314. *In*: Harrison, F. W., and R. R. Cowden (Eds.) Aspects of Sponge Biology. Academic Press, New York.

———. 1976. Life histories of coexisting Ceraclea species caddisflies (Trichoptera: Leptoceridae): the operation of independent functional units in a stream ecosystem. The Canadian Entomologist 108:1303–18.

Ricciardi A., and H. M. Reiswig. 1993. Freshwater sponges (Porifera, Spongillidae) of eastern Canada: Taxonomy, distribution, and ecology. Canadian Journal of Zoology 71:665–82.

Richelle-Maurer, E., Y. Degoudenne, G. van de Vyver, and L. Dejinghe. 1994. Some aspects of the ecology of Belgian freshwater sponges. In: van Soest, R. W. M., T. M. G. van Kempen, J. C. Brackman (Eds.). Sponges in time and space. Balkema, Rotterdam. p. 341–50.

Roback, S. S. 1968. Insects associated with the sponge *Spongilla fragilis* in the Savannah River. Notulae Naturae No. 412. The Academy of Natural Sciences of Philadelphia. p. 1–10.

Rollins, L. A. 1972. Poriferan Fauna of a Minnesota Pond. Journal of the Minnesota Academy of Science 38 (2): 83–85.

Rota, E., and R. Manconi. 2004. Taxonomy and ecology of sponge associate *Marionina* spp. (Clitellata, Enchytraeidae) from the Horomatangi Geothermal system of Lake Taupo, New Zealand. International Review of Hydrobiology 89:58–67.

Rothfuss, A. H., and J. S. Heilveil. 2018. Distribution of Sisyridae and freshwater sponges in the upper-Susquehanna Watershed, Otsego County, New York with a new locality for *Climacia areolaris* (Hagen). The American Midland Naturalist 180 (2): 298–305.

Ruengsawang, N., N. Sangpradub, T. Artchawakom, and R. Manconi. 2020. Aquatic insects in habitat-forming sponges: the case of the lower Mekong and conservation perspectives in a global context. Diversity 14(11):991.

Ruengsawang, N., N. Sangpradub, T. Artchawakom, R. Pronzato, and R. Manconi. 2017. Rare freshwater sponges of Australasia: new record of *Umborotula bogorensis* (Porifera: Spongillida: Spongillidae) from the Sakaerat Biosphere reserve in Northeast Thailand. European Journal of Taxonomy 260:1–24.

Ruperti, F., I. Belcher, A. Stokkermans, L. Wang, N. Marschlich, C. Potel, E. Maus, et al. 2024. Molecular profiling of sponge deflation reveals an ancient relaxant-inflammatory response.Current Biology 34:361-375.

Saller, U. 1988. Oogenesis and larval development of *Ephydatia fluviatilis* (Porifera, Spongillidae). Zoomorphology 108:23–28.

———. 1990. Formation and construction of asexual buds of the freshwater sponge *Radiospongilla cerebellata* (Porifera, Spongillidae). Zoomorphology 109:295–301.

Sand-Jensen, K., and M. F. Pedersen. 1994. Photosynthesis by symbiotic algae in the freshwater sponge, *Spongilla lacustris*. Limnology and Oceanography 39: 551–61.

Schindler, S., M. Wuttker, and M. Poschmann. 2008. Oldest record of freshwater sponges (Porifera: Spongillina) spiculite finds in the Permo-Carboniferous of Europe. Paläontologische Zeitschrift 82:373–84.

Schneider, P., and S. J. Hook. 2010. Space observations of inland water bodies show rapid surface warming since 1985, Geophysical Research Letters 37:L2245.

Schultz, D. T., S. H. D. Haddock, J. V. Bredeson, and R. E. Green. 2023. Ancient gene linkages support ctenophores as sister to other animals. Nature 618(7963):1–8.

Schuster A., S. Vargas, I. S. Knapp, S. A. Pomponi, J. R. Toonen, D. Erpenbeck, and G. Wörheide. (2018). Divergence times in demosponges (Porifera): first insights from new mitogenomes and the inclusion of fossils in a birth-death clock model. BMC Evolutionary Biology 18:114.

Scot, S. J., D. W. E. Nimmo, and L. M. Herrmann-Hoesing. 2019. New altitudinal records for, and ecomorphic variation in, two freshwater sponges (Porifera: Spongillida: *Spongilla lacustris* and *Ephydatia muelleri*) from a Colorado, USA, alpine lake. Western North American Naturalist 79:347–63.

Seigel, R. A., and R. J. Brauman. 1994. Food habits of the yellow-blotched map turtle (*Graptemys flavimaculata*). Museum Technical Report No. 28. Mississippi Museum of Natural Science. Mississippi Department of Wildlife, Fisheries and Parks, Jackson, MS. p. 1–19.

Simion, P., H. Phillippe, D. Baurain, M. Jager, D. J. Richter, A. Di Franco, B. Roure, et al. 2017. A large and consistent phylogenetic dataset supports sponges as the sister group to all other animals. Current Biology 27:958–67.

Skelton, J., and M. Strand. 2013. Trophic ecology of a freshwater sponge (*Spongilla lacustris*) revealed by stable isotope analysis. Hydrobiologia 107:227–35.

Smith, D. G. 1995. Keys to the freshwater macroinvertebrates of Massachusetts. Massachusetts Department of Environmental Quality Engineering, Division of water quality, Westborough, Massachusetts.

Smith, F. 1921. Distribution of the fresh-water sponges of North America. Illinois Department Registration Education Division Natural History Survey 14:10–22

———. 1922. A new locality for *Spongilla wagneri* Potts. Transactions American Microscopical Society 41:106.

Sowka, P. A. 1999. Occurrence of two species of freshwater sponges (*Dosilia radiospiculata* and *Ephydatia muelleri*) in Arizona. Southwest Naturalist 44:211–12.

Stein, B. A. 2002. States of the Union: ranking America's biodiversity. NatureServe, Arlington, VA. p. 1–25.

Stewart, B. A., P. G. Close, P. A. Cook, and P. M. Davies. 2013. Upper thermal tolerances of key taxonomic groups of stream invertebrates. Hydrobiologia 718:131–40.

Stoaks, R. D., J. K. Neel, and R. L. Post. 1983. Observations on North Dakota sponges (Haplosclerina: Spongillidae) and sisyrids (Neuroptera: Sisyridae). The Great Lakes Entomologist 16:171–76.

Strekal, T. A., and W. F. McDiffett. 1974. Factors affecting germination, growth, and distribution of the freshwater sponge, *Spongilla fragilis* Leidy (Porifera). The Biological Bulletin. 146:267–78.

Sublette, J. E. 1957. The ecology of the macroscopic bottom fauna in Lake Texoma (Denison reservoir), Oklahoma and Texas. American Midland Naturalist 57:371–402.

Sugden, S., J. Holert, E. Cardenas, W. Mohn, and L. Stein. 2022. Microbiome of the freshwater sponge *Ephydatia muelleri* shares compositional and functional similarities with those of marine sponges. The ISME Journal 16(11):1–10.

Talevska, A., B. Pejin, V. Kojic, T. Beric, and S. Stankovic. 2017. A contribution to pharmaceutical biology of freshwater sponges. Journal of Natural Product Letters 32:568–71.

Turner, E. C. 2021. Possible poriferan body fossils in early Neoproterozoic microbial reefs. Nature 596:87–91.

Van de Vyver, G., and Ph. Willenz. 1975. An experimental study of the life cycle of the

Freshwater sponge *Ephydatia fluviatilis* in its natural surroundings. Wihelm Roux's Archives 177:41–52.

Van Soest, R. W. M., N. Boury-Esnault, J. N. A. Hooper, K. Rützler, N. J. Voogd De, B. A lvarez Glasby, E. Hajdu, et al. 2017. *World Porifera Database*. Available online at http://www.marinespecies.org/porifera

Vaughn, C. M., and C. N. Brummel. 1963. The Spongillidae and Bryozoa of eastern South Dakota. South Dakota Academy of Science 42:103–7.

Vogel, S. 1974. Current induced flow through the sponge, *Halichondria*. Biological Bulletin (Woods Hole, Massachusetts) 147:41–52.

Volkmer-Ribeiro, C., H. L. Lenzi, F. Orefice, M. Peljo-Machado, L. M. de Alencar, C. F. Fonseca, T. CA Batista, et al. 2006. Freshwater sponge spicules: a new agent of ocular pathology, Memórias do Instituto Oswaldo Cruz 101 (8):899–903.

Volkmer-Ribeiro, C., and K. Rützler. 1997. *Pachyrotula*, a new genus of freshwater sponges from New Caledonia (Porifera: Spongillidae). Proceedings of the Biological Society of Washington 110 (4):489–501.

Waldschmidt, W. A., and L. W. LeRoy. 1944. Reconsideration of the Morrison Formation in the type area, Jefferson County, Colorado. Geological Society of America Bulletin 55:1097–114.

Wang, X., and W. E. G. Müller. 2011. Complex structures—Formation of siliceous spicules. Communicative and Integrative Biology 4 (6): 684–88

Watermolen, D. J. 2008. Catalog of North American state and regional freshwater sponge references. Misc. Pub. PUB-SS-1040, Bureau of Science Services, Wisconsin Department of Natural Resources. Madison, Wisconsin. p. 1–12.

Watts, T. M., M. Norris, and J. M. Piper. 2010. National Capital Region Network. Inventory and Monitoring Program a new record of *Ephydatia muelleri* in Prince William Forest Park (PRW). Natural Resource Technical Report NPS/NCRN/NRTR–2010/363. National Park Service, Fort Collins, Colorado.

Whitlock, H. N., and J. C. Morse. 1994. "*Ceraclea enodis*, a new species of the sponge-feeding caddisfly (Trichopteran: Leptoceridae) previously misidentified. Journal of the North American Benthological Society 13:580–91.

Wickham, M. M. 1922. Identification on fresh-water sponges in the Oklahoma fauna. Oklahoma Academy of Science 2:23–24.

Wielspütz, C., and U. Saller. 1990. The metamorphosis of the parenchymula–larva of *Ephydatia fluviatilis* (Porifera, Spongillidae). Zoomorphology 109:173–77.

Wilkinson, C. R. 1980. Nutrient translocation from green algal symbionts to the fresh water sponge *Ephydatia fluviatilis*. Hydrobiologia 75:241–50.

Wilkinson, C. R., and P. Fay. 1979. Nitrogen fixation in coral reef sponges with symbiotic cyanobacteria. Nature 279:527–29.

Williamson, C. W. 1979. Crayfish predation on freshwater sponges. American Midland Naturalist 101:245–46.

Woodward, G., D. M. Perkins, and L. E. Brown. 2010. Climate change and freshwater ecosystems: impacts across multiple levels of organization. Philosophical Transactions of the Royal Society 365:2093–106.

Wurtz, C. B. 1950. Fresh-water sponges of Pennsylvania and adjacent states. Notulae Naturae of the Academy of Natural Sciences 228:1–10.

Yin, Z., M. Zhu, E. H. Davidson, D. J. Bottjer, F. Zhao, and P. Tafforeau. 2015. Sponge grade body fossil with cellular resolution dating 60 Myr before the Cambrian. Proceedings of the National Academy of Sciences of the United States 112(12):1453–60.

INDEX OF COMMON AND SCIENTIFIC NAMES

Page numbers in **boldface** refer to species descriptions, photographs, and SEM images.

Acroneuria, 157
Actinobacteria, 51
Actinomycetes, 51
Actinopterygii, 2
alderflies, 52
algae, 49, 50, 58, 149, 161–66, 168
amphibians, 1, 10
amphipod, 150
Anheteromeyenia argyrosperma, 68, 169
Animalia, 168
Annelida, 10, 12
Arthropoda, 10, 12
Aythya collaris, 56

Baetis, 157
beetles, 52
blue-green algae, 164, 165
Brachycentrus, 157
Bryozoa/bryozoans, 24, 150, 152–54, 157

caddisfly, **45**, 54, 157
Calamites, 1
Ceraclea, **45**, 153
Ceraclea resurgens, 157
Ceraclea transversa, 54
Cherokeesia armata, **20**, 21, 67, 68, 73, **72–78**, 144, 150, 171, 184
Cheumatopsyche, 157
Chimarra, 157
chironomid larvae, 54, **55**, 56, 186
Chlorella, 49, 166, 168
Chordata, 10
Cladophora, 157
Climacia areolaris, 153, 155, 157, 159

Climacia chapini, 159
Cnidaria, **10**, 12, 24
Cocconeis, 157
Coelacanth, 2
Coleoptera, 52
Comamonas, 167
comb jellies, 10, 21
Cordaites, 1
Corophium lacustre, 150
Corvomeyenia carolinensis, 56, 68, 171, 179, 184
Corvomeyenia everetti, 49, 68, 171, 179, 181, 185
Corvospongilla becki, 56, 67, 68, 73, **81–86**, 144, 150, 171, 178, 181, 184
Corvospongilla lapidosa, 51
Corvospongilla novaeterrae, 68, 171, 179, 181
Corvospongilla siamensis, 54
crayfish, 18, 49, 66, 170
Cristatella mucedo, 152
Ctenophora/Ctenophores, 10–12, 189, 206
Cyanobacteria, 51, 52, 165
cyanosponges, 52
Cymbella, 157

Demospongiae, 66
Deutonymphs, 56
diatoms, 157
Diptera, 52, 54, 189
dobsonflies, 52
Dosilia palmeri, 68, 172, 178, 179, 181, 182, 184
Dosilia radiospiculata, 172, 178, 180–85
dragonflies, 52
Dreissena polymorpha, 53, 151
Duosclera mackayi, 172, 178, 179, 181–83, 185

Echinodermata, 10, 12
entoproct, 150
Ephemerella, 157
Ephemeroptera, 54, 165
Ephydatia fluviatilis, 3, 40, 51, 56, 58, 68, 73, **86–91**, 144, 151, 158, 172, 178–85
Ephydatia millsi, 68, 172, 179
Ephydatia muelleri, 3, 27, 68, 73, **91–96**, 145, 149, 152–54, 157, 166. 167, 173, 178–85
Ephydatia subtilis, 68, 179, 171
Equisetum, 1
Eunapius carteri, 51, 56
Eunapius fragilis, 3, 21, 55, 56, 68, 73, **96–100**, 144, 150–154, 157, 158, 173, 174, 178-185

fishflies, 52
Fragilaria, 157
fungi, 2, 3, 20, 49, 54, 166, 191

Gobius alcockii, 56
Gomphonema, 157
Graptemys flavimaculata, 56
green algae, 49, 50, 105, 124, 145, 166, 168

Halichondria bowerbanki, 166
Hantzschia, 157
Hastaperla, 157
Hemiptera, 52
Heptagenia, 157
Heteromeyenia baileyi, 68, 174, 179–85
Heteromeyenia latitenta, 68, 73, **101–5**, 144, 154, 174, 178, 180–84
Heteromeyenia longistylis, 68, 174, 183
Heteromeyenia riojai, 68
Heteromeyenia tentasperma, 68, 174, 182, 183, 185
Heteromeyenia tubisperma, 21, 68, 73, **105–10**, 144, 153, 174, 175, 179–85
Heterorotula lucasi, 21, **110–15**, 144, 155, 168, 173, 175, 184
Hippospongia, 2

Hydropsyche, 157
Hydroptila, 157

Isoperla, 157

Lepidodendron, 1
Lepidoptera, 52
Limnephilus, 157
lobe-finned fish, 2
Lubomirskia baikalensis, 166
Lubomirskiidae, 66, 190
lungfish, 2
Lycopods, 1

Malawispongiidae, 66, 191
mayflies, 52, 157, 165
Megaloptera, 52
Metaniidae, 66, 68, 71, 191
Metazoa, 10, 191
Metschnikowiidae, 66, 191
mites, 52, 56
Molanna, 157
Mollusca, 19, 12
moss animals, 24
moths, 53
mussels, 10, 18, 54, 151, 153, 188

Navicula, 157
Nematoda, 10
Nemoura, 157
Neuroptera, 52, 54, 191

Odonata, 52
Oncosclera kaniensis, 80
Orconectes, 55
Oscillatoria, 52

Palaeospongilla chubutensis, 15
Paleospongillidae, 66
Paludicella articulata, 153, 154, 157
Paragnetina, 157

Paraleptophlebia, 157
Parazoa, 10, 191
Pectinatella magnifica, 152, 153
periphyton, 168
Platyhelminthes, 10
Plecoptera, 52
Plumatella emarginata, 154
Porifera, 8, 10, 12, 66, 192
Potamolepidae, 21, 48, 66, 68, 71, 72, 80, 192
Potamophloios canadensis, 180
Pottsiela aspinosa, 68, 175, 179–83, 185
Pseudomonas, 151
Pteronarcys, 157
Pungu maclareni, 55

Racekiela biceps, 68
Racekiela cresciscrystae, 68
Racekiela montemflumina, 68
Racekiela pictouensis, 68
Racekiela ryderi, 21, 73, **115–20**, 145, 166, 174, 179–85
racoon, 49
Radiospongilla cerebellata, 43, 44, 68, 73, **120–24**, 143, 144, 149, 154, 156–85
Radiospongilla crateriformis, 21, 68, 73, **124–28**, 144, 156, 176, 178–85
ray-finned fishes, 2
Rhodospirillales, 167
Rhopalodia, 157
ring-necked duck, 56

Sarcopterygians, 2
Sarcopterygii, 2
Sediminibacterium, 167
Sigillaria, 1
Sisya vicaria, 151, 153, 155, 156
Sphenopsids, 1
Spirogyra, 157
Spongia, 2
Spongilla alba, 41, 58, 68, 176, 179, 181
Spongilla cenota, 68, 176, 179, 184

spongillafly, **52**, 151, 157, 159
Spongilla lacustris, 3, 8, 21, 29, 39, 45, 48, 51, 55, 58, 68, 73, **128–33**, 143, 144, 149, 151–58, 176–85
Spongilla leidyi, 67
Spongilla wagneri, 68, 177, 179, 181, 184
Spongillida, 65, 66, 68, 112, 143, 193
Spongillidae, 48, 66, 68, 71, 73, 80, 193
Staphylococcus, 3
Stenonema, 157
stoneflies, 52, 157
Stratospongilla penneyi, 68, 177, 179
Streptococcus, 3
Synechococcus, 166

Taeniopteryx, 157
Tanytarsus, 56
Trichoptera, 52, 54, 165, 193
Trochospongilla horrida, 56, 66, 68, 71, **133–27**, 144, 150, 177–84
Trochospongilla leidyi, 67, 68, 73, **137–41**, 144, 150, 177–84
Trochospongilla pennsylvanica, 67, 68, 178–85
Trombidiformes, 52
Tunicata, 34

Unionicola crassipes, 56
unionid mussel(s), 53, 153, 157
Urnatella gracile, 150

yellow-blotched map turtle, 56
yellow-green algae, 49

zebra mussel, **53**, 54, 151, 152, 154, 157, 159

The University of Tennessee Press
is a founding member of the
Association of University Presses.

Composed in 11/13.5 Arno Pro
with Acumin Pro display
by Kelly Gray
at the University of Tennessee Press.
Designed by Kelly Gray.

University of Tennessee Press
1015 Volunteer Blvd
Hodges Library 323
Knoxville, TN 37996-1000
www.utpress.org

www.ingramcontent.com/pod-product-compliance
Lightning Source LLC
Chambersburg PA
CBHW070244190225
22176CB00004B/192